Winifred G. Nayler

The Endothelins

With 69 Figures and 55 Tables

Springer-Verlag
Berlin Heidelberg New York London
Paris Tokyo Hong Kong Barcelona

Winifred G. Nayler, D.Sc.
The University of Melbourne
Department of Medicine
Austin and Repatriation Hospital
Heidelberg, Victoria, 3084
Australia

ISBN 3-540-52856-3 Springer-Verlag Berlin Heidelberg New York
ISBN 0-387-52856-3 Springer-Verlag New York Berlin Heidelberg

Printed: Kieser, 8902 Neusäß
Bookbinding: J. Schäffer, 6718 Grünstadt 1
2127/3140/543210 – Printed on acid-free paper

This book is dedicated to my husband, who above all else taught me the importance of the "three p's" – patience, perseverance and punctuality. I have conquered the latter two, but the first still requires some practise!

Preface

"What we call the beginning is often the end, and to make an end is to make a beginning. The end is where we start from."
T. S. ELIOT, "Geroniton"

The discovery of endothelin-1 was both an end and a beginning.

It is an end because this discovery has started a plethora of investigations as to endothelin's sites of production, mode of action, it's normal and abnormal physiological role and its involvement in a variety of pathological conditions ranging from renal failure to myocardial infarction, cardiac hypertrophy, hypertension, cyclosporine induced toxicity, mental depression, renal failure, stress related illness, and even stroke and sudden death.

It is a beginning because no one can foresee where the investigations will ultimately lead, or how they will end.

The discovery of the endothelins was a very recent event and soon specialist books will start to appear dealing in detail with different aspects of their chemistry, physiological role and pathological importance. The present book is intended to act as a guide and primer for those who are becoming intrigued by the many endothelin-related articles which are now appearing in the journals but who as yet have not had the time or opportunity to be caught up in the web of enthusiasm and excitement which is motivating those more directly involved in the field of research.

Accordingly, the first few chapters are intended to provide readers with an insight into the chemistry of the various endothelins. Then follow chapters dealing with their production and mode of action at a subcellular level, including their indirect modulation of the calcium channels. The remaining chapters draw attention to current speculation as to the possible involvement of the endothelins in a variety of pathological conditions, and the pharmacological interventions which may be used to moderate their activity.

Whilst the writing of any technical book or treatise brings its own rewards – in the sense of providing a golden opportunity to delve into the literature, to theorize and to categorize, such a task is seldom accomplished unaided. When preparing this book I have been fortunate in working with colleagues who share my keen interest in any topic which relates to calcium – and the endothelins most surely rely on calcium for their mode of action. Those colleagues include Dr. M. J. Daly, Dr. X. H. Gu and Dr. J. Liu, and I am grateful to them for their willingness to debate many of the issues which are discussed in the following chapters. In addition, and of equal importance, is the help given me by my secretaries, Reneé Janssen and Wendy Vanags, without whose help this book would have remained an illegible mass of pages. Diagrams are an essential component of most scientific texts and here, as in the past, I am grateful for the help so willingly given to me by Sianna Panagiotopoulos.

Above all else, I want to thank my husband for his unfailing support – although he prefers to remain anonymous!

pages. Diagrams are an essential component of most scientific texts and here, as in the past, I am grateful for the help so willingly given to me by Sianna Panagiotopoulos.

Above all else, I want to thank my husband for his unfailing support – although he prefers to remain anonymous!

Contents

Chapter 9 *Plasma, Tissue and Urine Endothelin-1*

Chapter 10 *Effect of Various Pathological Conditions on Plasma Endothelin*

Chapter 11 *Endothelin and the Kidney*

Chapter 12 *Endothelin and the Brain*

Chapter 1

The Discovery and General Properties of the Endothelins

"Begin at the beginning, and go on until you come to the end: then stop ..." from the king's advice to the White Rabbit.
LEWIS CAROL, in: "Alice in Wonderland."

The "endothelins" are potent vasoconstrictors (Yanagisawa et al. 1988a; Yanagisawa and Masaki 1989a, b). Although discovered only a few years ago, and then almost by chance, they are now providing a major challenge to workers in the field, particularly with respect to their site of production, mode of action, and involvement in physiological and pathological processes (Table 1.1). Some very basic questions are being asked, including whether these substances, which are peptides, are involved in the aetiology of hypertension, coronary and cerebral vasospasm, asthma, myocardial ischaemia, renal artery stenosis or even atherosclerosis. Questions are also being raised as to whether they contribute to the maintenance of basal homeostasis in the circulation, whether some of them function as neuropeptides, or whether they are nothing more than an evolutionary "hangover" – a question which has been raised because the toxins which make the venom of the burrowing asp so deadly have a chemistry and mode of action which is remarkably similar to that of the endothelins (Chapter 4). This similarity alone, however, does not provide sufficient grounds for assuming that the endothelins are simply an evolutionary remnant from the remote past, when the possession of such substances would have provided a useful means of overpowering prey, of hastening wound repair, or limiting haemorrhage.

What, then, are these substances which are now universally known as *"the endothelins"*, and how were they discovered?

Table 1.1. Some pathological conditions associated with altered levels of plasma endothelin-1

Condition	Response	Reference
Uraemia	Increase	Koyama et al. 1989
Essential hypertension	Increase	Saito et al. 1990
Myocardial infarction	Increase	Miyauchi et al. 1989a
Cardiogenic shock	Increase	Cernacek and Stewart 1989
Surgery	Increase	Hirata et al. 1989b
Endotoxic shock	Increase	Pernow et al. 1989a
Pulmonary hypertension	Increase	Cernacek and Stewart 1989

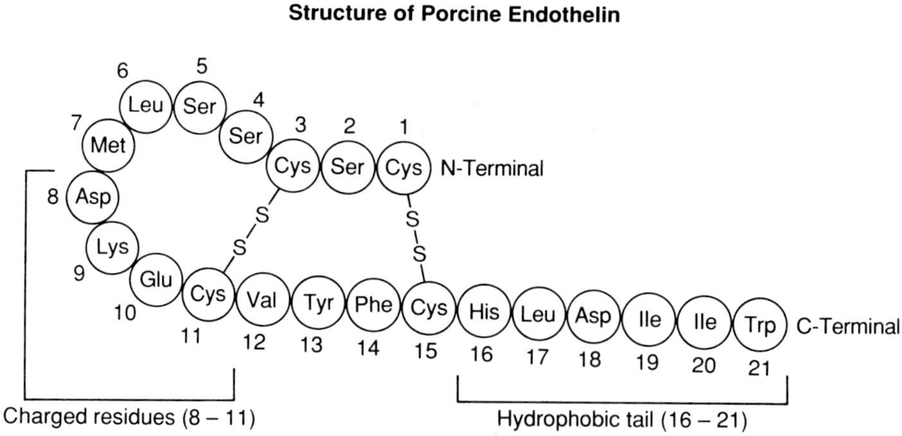

Fig. 1.1. Chemical structure of endothelin-1. Note the presence of twenty-one amino acids, of two disulphide bridges (between residues *1* and 15, and *3* and *11*), of a hydrophobic tail and a terminal tryptophan residue

The Endothelins

The endothelins are polypeptides. Each member of the group contains:
(i) twenty-one amino acids;
(ii) two intrachain disulphide bridges (which occur at fixed positions between residues *1* and *15*, and *3* and *11*);
(iii) a hydrophobic tail, and
(iv) a terminal tryptophan residue.

These essential chemical features – the twenty-one amino acids, the intrachain disulphide bridges, the hydrophobic tail and the terminal tryptophan residue – are readily identifiable in the chemical structure of the prototype, endothelin-1 (ET-1), shown in Figure 1.1. This particular endothelin was originally purified by Yanagisawa and his co-workers (Yanagisawa et al. 1988a) from the supernatant of cultured porcine aortic endothelial cells. It was these same investigators who coined the name, *endothelin* – a logical choice at the time. Once *endothelin* (or endothelin-1, or ET-1 as it is now known as) was sequenced and cloned interest in it intensified, mainly because of its ability to trigger *sustained* constriction even when administered in minute amounts. It did not take long to establish that this constrictor response involved Ca^{2+} (Simonson and Dunn 1990).

The first endothelin to be purified came from the medium bathing cultured porcine aortic endothelial cells – hence its colloquial name, *porcine endothelin*. However, with the benefit of hindsight, it is now apparent that this particular endothelin is only one of a family of chemically homologous, biologically active polypeptides. Others (Table 1.2) include:

Table 1.2. The endothelins

Endothelin-isoforms	
Endothelin-1 (ET-1, or porcine ET)	
Endothelin-2[a] (ET-2)	or Trp^6 Leu^7 ET-1
Endothelin-3[a,b] (ET-3, or rat ET)	or Thr^2 Phe^4 Thr^5 Tyr^6 Lys^7 Tyr^{14} ET-1
VIC[c] (mouse β endothelin)	or Asn^4 Trp^6 Leu^7 ET-1

[a] Yanagisawa and Masaki 1989a, b
[b] Yanagisawa et al. 1988b
[c] Saida et al. 1989

(a) endothelin-2 (ET-2);
(b) endothelin-3 (ET-3 or rat ET); and
(c) VIC (or "mouse" β endothelin).

The chemical structure of each of these isoforms is described in the next chapter. However, even at this stage it is worth noting that although each polypeptide retains the essential chemical characteristics of the group – the twenty-one amino acids, the two pairs of disulphide linkages, the hydrophobic tail and the terminal tryptophan residue – minor differences in amino acid composition, sometimes only involving two amino acids, results in marked changes in their physiological activity. The following two examples serve to illustrate the point. Firstly, with respect to their vasoconstrictor potency, endothelin-1 > endothelin-2 > endothelin-3 (Inoue et al. 1989a; Matsumoto et al. 1989). Secondly, although VIC differs from endothelin-1 in only three amino acid residues (Table 1.2) it's constrictor effect on ileal smooth muscle greatly exceeds that of endothelin-1 (Saida et al. 1989). These particular differences are mentioned here only to illustrate the fact that although chemically very similar, the various endothelins have different physiological profiles. As a corollary, it is possible that the individual isoforms have different physiological functions. They may even be involved in different pathological conditions.

The Discovery of the Endothelins

Readers of the recent endothelin literature may quite reasonably have assumed that the discovery of these polypeptides started with the chemical sequencing and cloning of endothelin-1 (Yanagisawa et al. 1988a). Certainly the sequencing and cloning of the polypeptide was a major achievement, and of immense importance to the field, as was the identification of its precursors. However, the paper in which this information was presented was actually preceded by three other highly relevant papers, and one abstract (Table 1.3). The abstract, which was the first to appear, was published by O'Brien and McMurtry in 1984 (O'Brien and McMurtry 1984). In that abstract they reported that pulmonary arterial and aortic endothelial cells secrete a polypeptide which causes sustained constriction in bovine pulmonary arteries. In the following year a paper by Hickey, Rubanyi, Paul and Highsmith (Hickey et al.

Table 1.3. Important dates associated with the discovery of the endothelin family of polypeptides

1984	O'Brien and McMurtry obtained vasoconstrictor polypeptide from pulmonary artery and endothelial cells.
1985	Hickey et al. obtained vasoconstrictor substance from the supernatant of bovine aortic endothelial cells.
1986	Gillespie et al. obtained vasoconstrictor substance from the supernatant of cultured bovine aortic endothelial cells and noted its long-lasting effect and relative selectivity for the vasculature.
1987	O'Brien et al. confirmed the polypeptide nature of the vasoconstrictor substance and estimated its molecular weight to be around 3000 daltons. They noted blunting of the constrictor response when verapamil was added.
1988	Yanagisawa et al. isolated and sequenced the polypeptide, *named it "endothelin"*, and cloned its precursor, mRNA-preproendothelin.
1989	Other isoforms identified, including endothelin-2 (ET-2) and endothelin-3 (ET-3) (Inoue et al. 1989a) and VIC (Saida et al. 1989).

1985) appeared in which they described the properties of a potent vasoconstrictor substance found in the supernatant of cultured *bovine* aortic endothelial cells. After noting that this substance was protease-sensitive, and that it caused a dose-dependent constrictor response in porcine, bovine and canine coronary arteries, *provided that extracellular* Ca^{2+} was available, Hickey et al. (1985) concluded that they were dealing with an endothelial-derived vasoconstrictor polypeptide which might be *"important in the regulation of vascular smooth muscle contraction"*. At the same time they established that a wide variety of antagonists, including antagonists of the α and β adrenergic, serotonergic, histaminergic and cholinergic receptor systems, as well as cyclooxygenase and lipoxygenase inhibitors, failed to suppress the constrictor response.

In the following year (*1986*) another relevant paper appeared. This time the authors were Gillespie, Owasoyo, McMurtry and O'Brien (Gillespie et al. 1986). These investigators had been studying the vasoconstrictor properties of a substance they found to be secreted by cultured bovine aortic endothelial cells and which they named *endothelial-derived constrictor factor* (or EDCF). The coronary constrictor effect of this substance (EDCF) was dose-dependent, sustained, and occurred at a concentration which had no effect on either left ventricular function or heart rate. Gillespie and his colleagues ruled out any involvement of either angiotensin or substance P and concluded that the substance was a peptide released from endothelial cells and capable of provoking *"sustained coronary vasoconstriction by an unknown mechanism"*.

The next year (*1987*), O'Brien, Robbins and McMurtry published another paper on this topic (O'Brien et al. 1987) in which they further characterized the properties of the constrictor substance which was being secreted by bovine aortic and pulmonary artery endothelial cells. They confirmed it's polypeptide nature, established that it had a molecular weight of around 3000 daltons, and agreed that its constrictor effect was sustained and insensitive to serotoninergic, histaminergic, α adrenergic, opoid, leukotriene, angiotensin II, or substance P receptor antagonists. However, they did note that its constrictor effect was *partially* reversed by the calcium antagonist, verapamil, and *completely* overcome by nitroprusside and isoprenaline. They went on to establish that although the substance was not extractable into organic solvents its activity was completely destroyed by trypsin and by incubation with a neutral protease.

Hence, by mid-1987 several groups had already shown that aortic endothelial cells secrete a vasoactive polypeptide, the constrictor effect of which is sustained, but can be completely overcome by either isoprenaline or nitroprusside. The year after the paper by O'Brien et al. (1987) appeared, Yanagisawa and his colleagues (Yanagisawa et al. 1988a) succeeded in sequencing this hitherto evasive polypeptide. At the same time they named it *endothelin* and extended observations relating to its *sustained* constrictor effect by demonstrating sustained vasoconstriction under *in vivo* conditions. They confirmed its remarkable potency as a vasoconstrictor agent, noting that its potency even exceeded that of angiotensin II, vasopressin or neuropeptide Y, making it the most potent mammalian vasoconstrictor polypeptide to be identified so far. In total agreement with O'Brien and his colleagues, Yanagisawa's group (Yanagisawa et al. 1988a) found that the coronary constrictor effect of this polypeptide was completely overcome by isoprenaline. They also confirmed Hickey's earlier observation relating to the Ca^{2+}-dependence of the coronary constrictor response and ended their paper by concluding that "an influx of Ca^{2+} is required for the action of endothelin".

To dismiss the importance of this paper by Yanagisawa and his colleagues (Yanagisawa et al. 1988a) as mere confirmation of pre-existing data would be highly irresponsible, since that paper contained *the vital data concerning the structure, synthesis and mode of action of the polypeptide.* Yanagisawa and his colleagues even identified many of the conditions (thrombin, Ca^{2+} ionophore A27187, adrenaline and shear stress) which stimulate it's production. From the moment Yanagisawa's group published their findings – only two years ago – the field has attracted a frenzy of interest, resulting in many surprising and tantalizing discoveries and unanswered questions. For instance why should plasma endothelin-1 levels be abnormally high in patients with essential hypertension (Saito et al. 1990), renal failure (Koyama et al. 1989), myocardial ischaemia (Miyauchi et al. 1989a) and cardiogenic shock (Cernacek and Stewart, 1989)? Why do the plasma levels rise during surgery (Hirata et al. 1989b), during subarachnoid haemorrhage (Masaoka et al. 1989) and during endotoxic shock (Pernow et al. 1989a) and why is the level abnormally high in bronchial exudates of patients with asthma (Nomura et al. 1989) and low in the cerebrospinal fluid of mentally depressed patients (Hoffman et al. 1989)? In each case the question must be asked as to whether these changes are the cause of or secondary to the underlying pathology. Or are they irrelevant?

General Properties of Endothelin-1

The chemistry of the endothelins, the steps involved in their synthesis and the properties of the chemically related toxins are discussed in the following chapters

Table 1.4. Properties of endothelin-1

	Response	Reference
A. Tissue		
1. Blood vessels		
(i) Coronary arteries	Constriction	Chester et al. 1989 Tippins et al. 1989
(ii) Renal arteries	Constriction	King et al. 1989
(iii) Cerebral arteries	Constriction	Asano et al. 1989
(iv) Microvasculature	Constriction	Brain et al. 1988
(v) Veins	Constriction	Miller et al. 1989a
(vi) Iris sphincter muscle	Contraction	Geppetti et al. 1989
2. Lymphatics	Constriction	Fortes et al. 1989
3. Airways		
(i) Trachea	Constriction	Maggi et al. 1989a
(ii) Bronchus	Constriction	Uchida et al. 1988
4. Heart	Positive inotropy	Ishikawa et al. 1988
5. Brain	Behaviourial disorders Pressor response Neurosecretion (?)	Moser and Pelton 1989 Ouchi et al. 1989 Yoshizawa et al. 1990
6. Nerves	Depolarization	Yoshizawa et al. 1989b
7. Stomach	Ulcerogenic	Wallace et al. 1988
8. Platelets (ex vivo)	Anti-aggregatory	Thiermermann et al. 1988
9. Cells	Proliferation	Hirata et al. 1989d Simonson et al. 1989
10. Ileum	Constriction	Hiley et al. 1989
B. Biochemical		
(i) Cytosolic Ca^{2+}	Increased	Ohnishi et al. 1989
(ii) Na^+:H^+ exchange	Increased	Simonson et al. 1989
(iii) Phospholipase C	Increased	Simonson et al. 1989
(iv) Phosphoinositol metabolism	Increased	Marsden et al. 1989
(v) Protein kinase C activity	Increased	Marsden et al. 1989
(vi) Diacylglycerol accumulation	Increased	Griendling et al. 1989
C. Direct effect on release of other substances		
(i) Vasopressin	Increased	Schichiri et al. 1989
(ii) ANP	Increased	Fukuda et al. 1988
(iii) EDRF	Increased	Warner et al. 1988, 89
(iv) Noradrenaline	Increased	Miller et al. 1988
(v) Prostacyclin	Increased	de Nucci et al. 1988
(vi) Renin	Reduced	Matsumura et al. 1989
D. Others		
Conversion of Angiotensin I to II	Increased	Kawaguchi et al. 1990

This list is not meant to include all of the known activities of endothelin-1. It is included here simply to provide some idea the broad spectrum of responses.

* The vasoconstrictor effect of endothelin-1 is often preceded by a transitory dilator response.

ANP (atrionatriuretic peptide); EDRF (endothelial-derived relaxing factor)

(Chapters 2–4). As a prelude to these chapters, however, it may be worth considering some of the physiological and biochemical systems which they influence. This applies particularly to endothelin-1 (Table 1.4). The other endothelins may be just as active, but as yet information relating to their activities is limited, primarily because of their more recent discovery.

Even a cursory consideration of the information summarized in Table 1.4 provides some idea of endothelin-1's broad spectrum of activity. Thus, apart from widespread arterial vasoconstriction, it causes bronchoconstriction (Uchida et al. 1988; Maggi et al. 1989a), and constricts venous and lymphatic vessels (Miller et al. 1989a; Fortes et al. 1989). It depolarizes spinal neurones (Yoshizawa et al. 1989a, b) and has a positive inotropic and chronotropic effect on the heart (Ishikawa et al. 1988a, b). It activates the phosphoinositide system (Marsden et al. 1989a; Simonson and Dunn 1990), mobilizes intracellular Ca^{2+} (Ohnishi et al. 1989), is mitogenic (Simonson et al. 1989; Hirata et al. 1989d), stimulates diacylglycerol accumulation, activates protein kinase C (Griendling et al. 1989), stimulates phospholipase C (Resnik et al. 1989), and inhibits platelet aggregation *ex vivo* (Thiermermann et al. 1988). It inhibits renin release (Matsumura et al. 1989) but stimulates the secretion of vasopressin (Shichiri et al. 1989), noradrenaline (Miller et al. 1988), atrionatriuretic peptide (Fukuda et al. 1988), endothelial-derived relaxing factor (EDRF) (Warner et al. 1988, 1989) and prostacyclin (de Nucci et al. 1988). This is not the end of the list, however, because endothelin-1 has many other properties, including an ability to activate the Na^+:H^+ exchanger (Simonson et al. 1989) and, even in minute doses, to contract afferent and efferent renal arteries (King et al. 1989).

Having regard to the probable involvement of endothelin-1 in such a wide variety of events and physiological and pathological conditions it might reasonably have been assumed that this, and the chemically homologous polypeptides, should have been discovered along with the other vasoactive polypeptides, vasopressin and angiotensin. Even when the endothelins were identified it can be questioned why initial interest in them centred around their potency as vasoconstrictor agents? Hours could be spent answering and debating these questions, but when reduced to their simplest form they can be answered along the following lines:

(i) until the endothelin-derived relaxing factor (EDRF) was discovered and characterized, little attention was given to the secretory potential of the endothelium;
(ii) some time elapsed before non-endothelial sites of endothelin production were identified;
(iii) time was needed to develop an assay capable of detecting the relatively low endothelin levels which are present in various tissues, including plasma, urine, lymph, cerebrospinal fluid, bronchial exudates, human milk, brain and kidney, and
(iv) studies relating to the mode of action of these polypeptides were largely delayed until supplies of synthetic endothelin became available and the receptors were identified (Hirata et al. 1988; 1989a).

Once these various difficulties were overcome, however, information concerning the mode of action, sites of production and physiological and pathological significance of these polypeptides began to emerge. At the same time it became apparent that they increase cytosolic Ca^{2+} (Ohnishi et al. 1989; Fu et al. 1989).

Summary

The search for an endothelial-derived vasoconstrictor substance can be traced back to 1984. Within four years of the first report appearing on this subject the molecule and its isoforms have been identified and sequenced, their activity documented and the biosynthetic pathways involved in their production established. Since 1988 the field has developed at an amazingly rapid rate. Human cultured endothelial cells have been shown to secrete endothelin-1 (Clozel and Fischli 1989), and specific receptors have been located in a wide variety of tissues – including not only the vasculature but also the heart, brain, lungs, placenta, uterus, kidneys, trachea, dorsal root neurones and stomach. Endothelin-1 has been found to be present in blood, cerebrospinal and lymphatic fluids, bronchial exudates, human milk and urine, with the levels sometimes changing under certain pathological conditions. There is a very real possibility of these polypeptides being involved not only in the normal maintenance of cellular function but in addition, in various disease states (Nayler 1989, 1990) – including cardiogenic shock, hypertension, mental depression and myocardial ischaemia.

The prototype of the group, endothelin-1, has a wide spectrum of activity. It raises intracellular Ca^{2+}, activates the phosphoinositol system, releases vasopressin, noradrenaline, prostacyclin, endothelial-derived relaxing factor (EDRF) and the atrionatriuretic peptide (ANP). It inhibits renin release, causes cell proliferation (and therefore may be atherogenic), is positively inotropic and constricts veins, lymphatics and bronchiolar, uterine and tracheal smooth muscle, as well as arteries. It is also ulcerogenic, affects platelet aggregation and can cause behaviourial disturbances. Any molecule with such a broad spectrum of activity must surely warrant careful study. The search for a specific antagonist or antibody is a logical sequel.

As a prelude to discussing the physiological and pathological significance of the endothelins, and their mode of action, some attempt should be made to summarize their chemical characteristics, the steps involved in their biosynthesis, and the chemistry of some closely related toxins. These are the main topics of the next few chapters.

Chapter 2

The Chemistry of the Endothelins

"Everything must be like something: so what is this like?"
E. M. Forster, in: Our Diversions

Readers might be forgiven for asking what prompted O'Brien and McMurtry (1984, Table 1.1) to search for an endothelial-derived constrictor substance at a time when most investigators were busy following up Furchgott and Zawadzki's discovery relating to the ability of vascular endothelial cells to release the potent vasodilator *endothelial-derived relaxing factor*, (EDRF) (Furchgott and Zawadzki 1980). Moreover only a few years had elapsed since Moncada and his colleagues (Moncada et al. 1976) had discovered prostacyclin – the potent vasoactive substance released by arteries. It should be remembered, however, that these discoveries – and particularly those of Furchgott and Zawadzki – had alerted physiologists to the importance of the hitherto neglected endothelial cells. Until that time these thin, sheet-like cells (Figure 2.1) had attracted little attention, although their presence

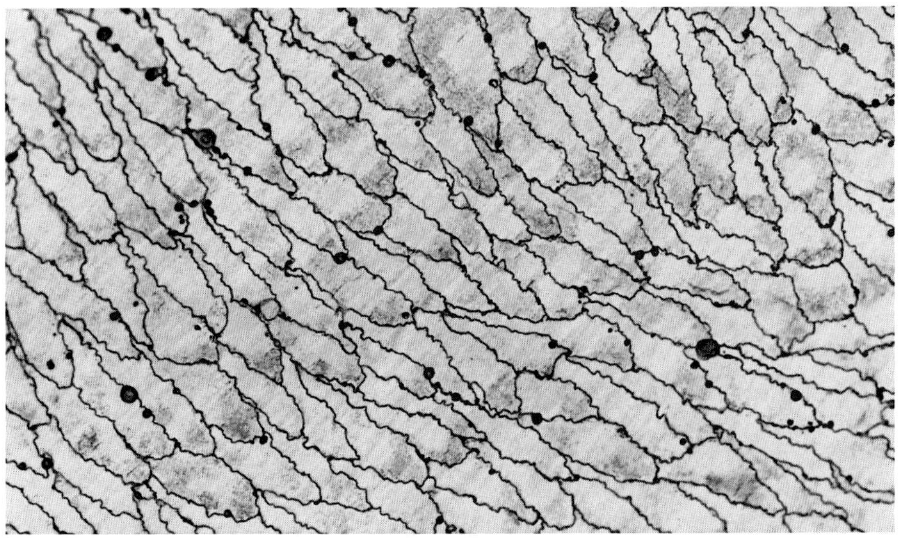

Fig. 2.1. A light micrograph of "an enface" preparation of dog femoral artery stained with silver nitrate. The endothelial cells comprise a continuous layer of cells. In the micrograph the borders are clearly outlined by the deposition of the silver stain (Mag X 430). Reproduced with permission by Dr. J. Mandenson and Dr. G. Campbell (J. Physiol.; 344:209–222, 1983)

Table 2.1. Tissues which secrete endothelins

Endothelin	Tissue	Reference
Endothelin-1	Vascular endothelial cells	Yanagisawa et al. 1988a Emori et al. 1989a, b Sawamura et al. 1989
	Kidney cell	Kosaka et al. 1989
	Spinal cord	Shinmi et al. 1989a
	Tracheal epithelial cells	Black et al. 1989 Rozengurt et al. 1990
	Retinal epithelial cells	Takahashi et al. 1989
	Dorsal root ganglia	Giaid et al. 1989
	Tumour cells	Suzuki et al. 1989
Endothelin-2	Kidney cells	Itoh et al. 1989
Endothelin-3	Intestine, pituitary and brain (rat)	Matsumoto et al. 1989
	Brain (porcine)	Shinmi et al. 1989b
VIC *(mouse β endothelin)*	Intestine	Saida et al. 1989

was well known and appreciated by histologists. This situation changed quite dramatically as soon as investigators began probing the properties of prostacyclin and EDRF. Perhaps O'Brien and McMurtry (1984) and Hickey and her colleagues (Hickey et al. 1985) recalled the YIN-YAN hypothesis of "equal and opposites" which had been so popular in the preceeding decade, particularly with those who were involved in nucleotide research. They could have argued that if endothelial cells release a powerful relaxing factor (EDRF and prostacyclin) then these cells might also secrete constrictor substances – as indeed they do (Yanagisawa et al. 1988a).

This chapter provides a brief account of the chemistry of endothelin-1 and its isoforms, *endothelin-2, endothelin-3 and VIC.* Of all these endothelins only endothelin-1 is known to be secreted by vascular endothelial cells, and even then its secretion is not limited to these cells. Thus, as can be seen from Table 2.1, kidney cells (Kosaka et al. 1989), tracheal epithelial cells (Black et al. 1989), renal glomeruli (Marsden et al. 1989b), motor and dorsal horn neurones of the spinal cord (Shinmi et al. 1989a), retinal endothelial cells (Takahashi et al. 1989) and even certain tumour cells (Suzuki et al. 1989) are capable of secreting this particular endothelin.

Endothelin-1 (or Porcine Endothelin, or ET-1)

Endothelin-1 is a potent vasoconstrictor polypeptide (Yanagisawa et al. 1988a). It has a molecular weight (Table 2.2) of 2491.8 and in common with the other isoforms is an acidic, twenty-one amino acid polypeptide (Figure 2.2), with two intrachain

Table 2.2. Molecular weights of the endothelins

Endothelins	Molecular weight (daltons)
Endothelin-1 (porcine endothelin, ET-1)	2491.8
Endothelin-2 (human endothelin, ET-2)	2547.5
Endothelin-3 (rat endothelin, ET-3)	2643.6
VIC (mouse β endothelin)	2574.5

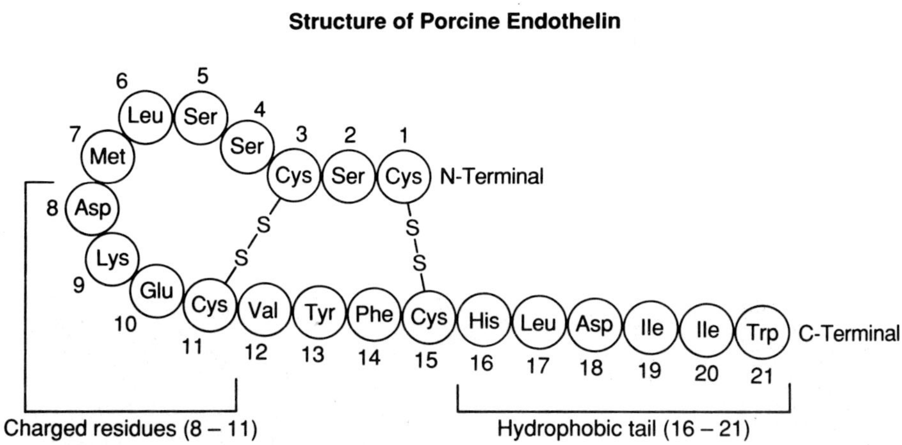

Fig. 2.2. Chemical structure of endothelin-1. (See Table 2.3 for the key to the amino acids)

disulphide bonds and a terminal tryptophan residue. It contains four cysteine residues (at fixed positions of *1, 3, 11* and *15*), and disulphide bonds link residues *1* and *15*, and *3* and *11* (Figure 2.2). There is a cluster of charged residues at positions *8 – 10* (aspartic acid – lysine – and glutamic acid) and a hydrophobic hexapeptide tail (residues *16 – 21*) terminating with a tryptophan residue. Without that terminal tryptophan residue the molecule loses most of its activity (Chapter 7). The configuration of the molecule as determined by nuclear magnetic resonance is shown in Figure 2.3 A–D. Clearly the molecule does have a head and a terminal tail (Figure 2.3 A and B) and is surrounded by van der Waal forces (Figure 2.3 C). Figure 2.3 D, which shows the water sheath, is of interest because it shows that water molecules cannot closely approach the molecule.

Endothelin-2 (ET-2)

As far as its chemistry is concerned endothelin-2 closely resembles endothelin-1 (Figure 2.4). It has a molecular weight of 2547.5 (Table 2.2). The two intrachain disulphide linkages are retained, as is the terminal tryptophan residue. However two

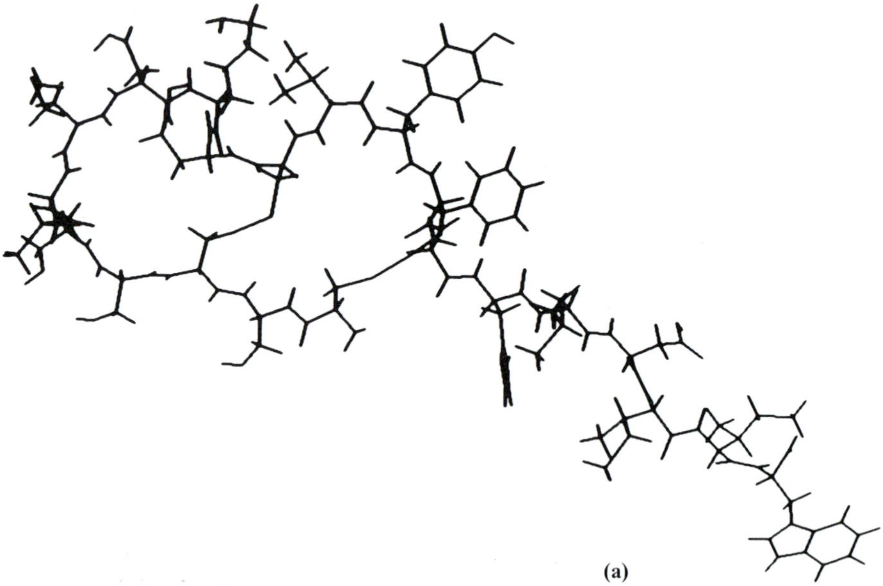

(a)

(b)

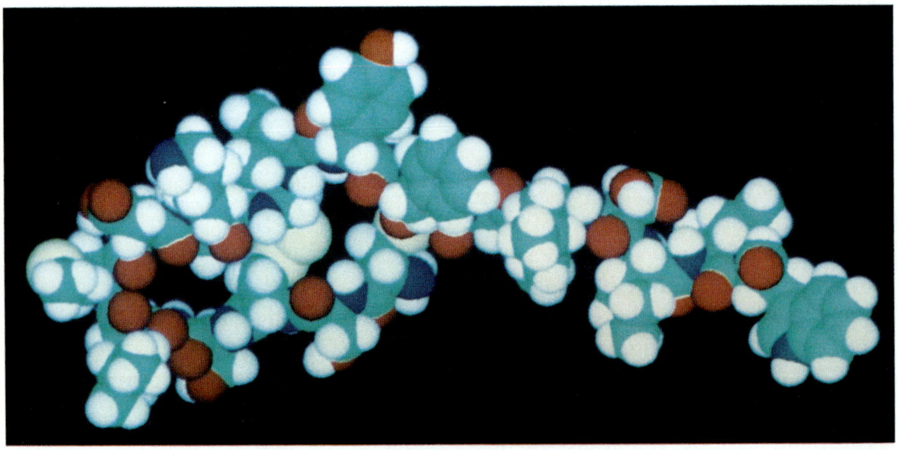

Fig. 2.3. NMR-determined structure of endothelin: (a) traditional stick diagram, (b) space filling model, (c) van der Waals surface, (d) water-accessible surface – denoted by outer sheath. (Reproduced with permission from Munro et al. 1990)

amino acid substitutions (Table 1.2) have occurred – at positions *6* and 7. In endothelin-1 these positions are occupied by leucine and methionine respectively. In endothelin-2 position *6* is occupied by tryptophan, and position *7* by leucine. Hence endothelin-2 (ET-2) is often described as Trp^6 Leu^7 endothelin-1 (or Trp^6 Leu^7 ET-1).

As far as the physiological properties of endothelin-2 are concerned its potency as a vasoconstrictor agent exceeds that of endothelin-1 (Inoue et al. 1989a).

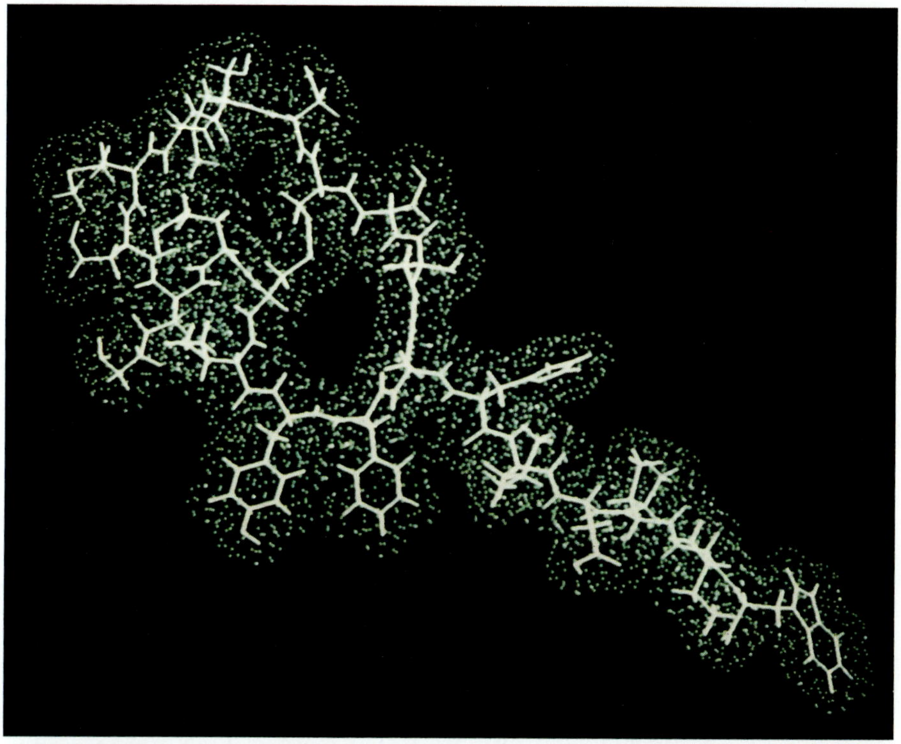

(c)

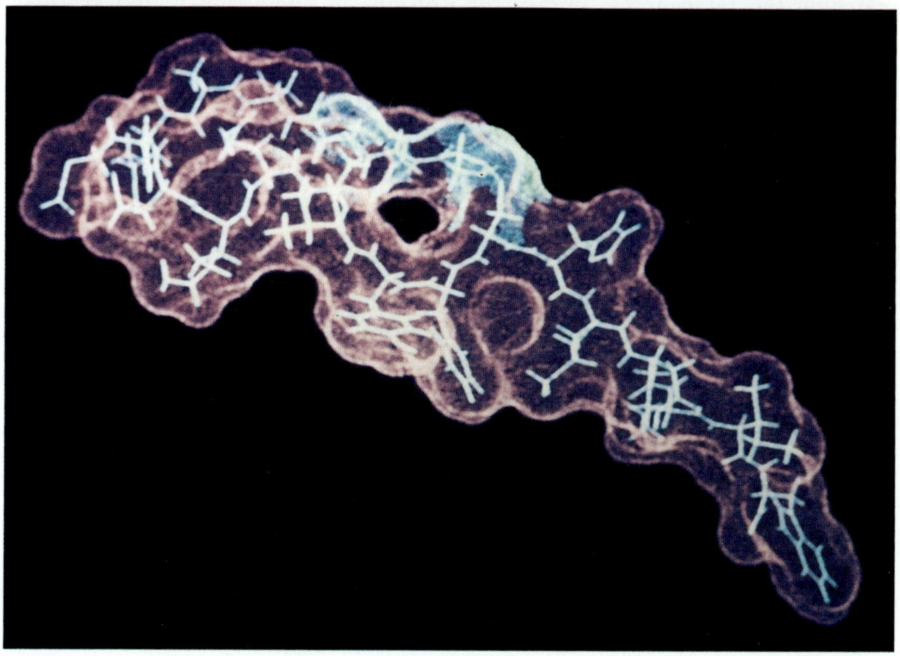

(d)

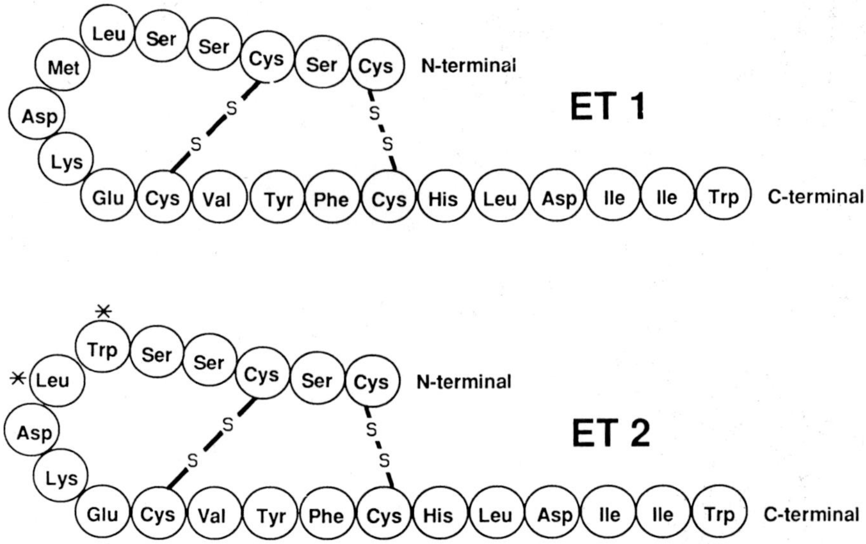

Fig. 2.4. Comparison between the chemical structures of endothelin-1, and endothelin-2. Note the location of the amino acid substitutions, marked with asteriks. (The key to the amino acids is shown in Table 2.3, along with their single letter coding)

Yanagisawa et al. (1988b) originally detected the endothelin-2 sequence when searching the rat genomic library for isoforms of endothelin-1, and synthetic endothelin-2 has been used to characterize its properties. Recently, however, Itoh et al. (1989) reported that this endothelin isoform is secreted by several kidney cell lines (Table 2.1). Hence it can no longer be regarded as being just a synthetic analogue of endothelin-1!

Endothelin-3 (Rat Endothelin, or ET-3)

This isoform has a molecular weight of 2643.6 (Table 2.2). As with endothelin-2, endothelin-3 was initially discovered by Yanagisawa and his colleagues (Yanagisawa et al. 1988b) when they were searching the rat genomic library for amino acid sequences resembling endothelin-1. It took less than one year for this isoform to be detected in a variety of tissues, including (Table 2.1) brain, pituitary gland and intestine (Shinmi et al. 1989b; Matsumoto et al. 1989).

Endothelin-3 differs from endothelin-1 to the extent of there being six amino acid substitutions (Figure 2.5). The substitutions occur at residues *2, 4, 5, 6, 7* and *14*. Residue *2* is occupied by threonine (instead of serine), residue *4* by phenylalanine (instead of serine), residue *5* by threonine (instead of serine), residue *6* by tyrosine (instead of leucine), residue *7* by lysine (instead of methionine) and residue *14* by tyrosine (instead of phenylalanine). Hence endothelin-3 (ET-3) is often described as

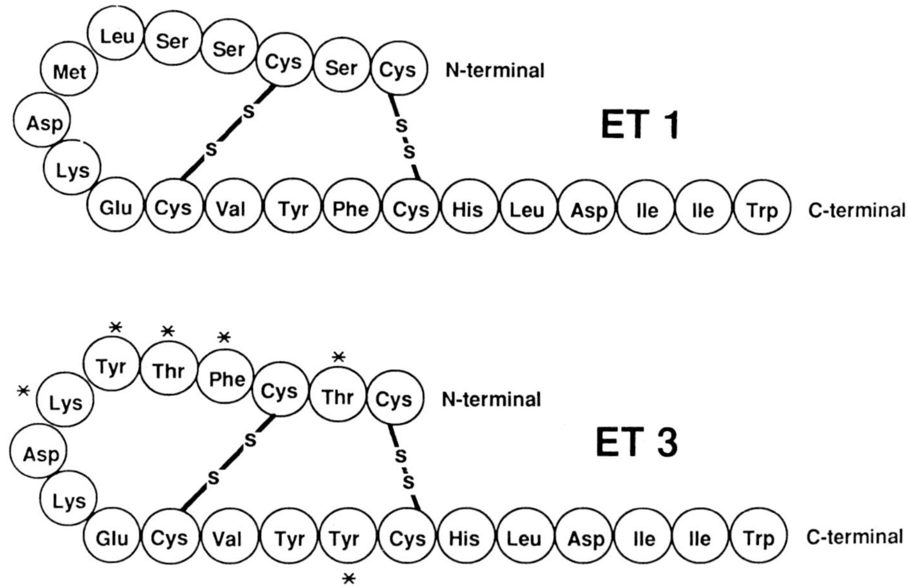

Fig. 2.5. Comparison between the chemical structure of endothelin-1, and endothelin-3. The chemical substitutions relative to endothelin-1 are marked with an asterisk. (The key to the amino acids is given in Table 2.3, along with their single letter coding)

being $Thr^2\ Phe^4\ Thr^5\ Tyr^6\ Lys^7\ Tyr^{14}$ endothelin-1 (or $Thr^2\ Phe^4\ Thr^5\ Tyr^6\ Lys^7\ Tyr^{14}$ ET-1). However:

(i) the twenty-one amino acid "backbone" of the molecule is retained;
(ii) the disulphide bridges between the cysteine residues are still present; and
(iii) the terminal tryptophan residue remains part of the terminal carboxy hexapeptide (histidine, leucine, aspartic acid, isoleucine, and tryptophan).

The substitution of the charged lysine residue for the methionine at position *7* of endothelin-1 means that endothelin-3 is more polar than endothelin-1. This may explain why endothelin-3 dissociates more readily than endothelin-1 from its binding sites on cell membranes.

Various roles have been assigned to endothelin-3. It exhibits fibrinolytic activity (Korbut et al. 1989), is an antiaggregatory agent under *in vivo* conditions and at dose levels which have minimal effects on blood pressure or heart rate (Lidbury et al. 1989). As far as its pressor activity is concerned, its constrictor activity is equivalent to that of angiotensin II, but only one twentieth that of endothelin-1 (Yanagisawa et al. 1989b). Probably, therefore, its primary role is not concerned with maintaining vascular tone. On the other hand it's presence in the brain has aroused speculation as to whether it might function as a neuropeptide.

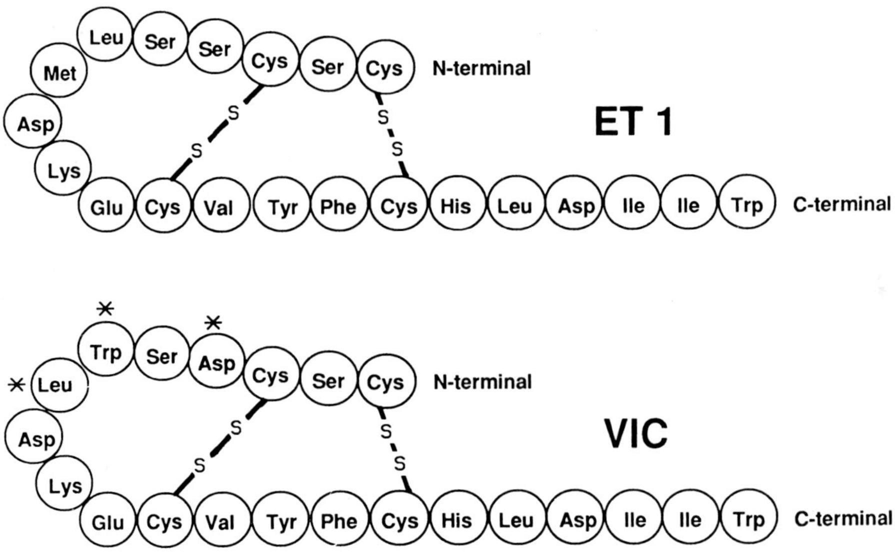

Fig. 2.6. Comparison between the chemical structure of endothelin-1, and that of VIC (mouse β endothelin). The chemical substitutions relative to endothelin-1 are marked with an asterisk

VIC [(Mouse) β Endothelin]

This is a novel polypeptide with a molecular weight (Table 2.2) of 2574.5, and which differs from endothelin-1 in only three amino acid residues, at positions *4, 6* and *7* (Figure 2.6). Residue *4* becomes asparagine, instead of serine, residue *6* becomes tryptophan instead of leucine, and residue *7* is leucine, instead of methionine. Hence VIC is Asp^4 Trp^6 Leu^7 endothelin-1 (Figure 2.6).

One of the interesting things about VIC is that although it has a pressor effect similar to that of endothelin-1, it is expressed in the intestine, and not in endothelial cells (Saida et al. 1989). It is a potent constrictor of ileal smooth muscle (Saida et al. 1989). Maybe, therefore, it functions as a gastrointestinal "hormone".

Nomenclature

Before proceeding any further with the endothelin story it may be useful to explain why the nomenclature relating to these polypeptides became so complex. For example, why is endothelin-1 sometimes called porcine endothelin? Why is endothelin-2 often referred to as human endothelin? And why is endothelin-3 also known as *rat* endothelin?

Endothelin-1

Because endothelin-1 was first isolated from the supernatant of porcine aortic endothelial cells (Yanagisawa et al. 1988a) it initially became known as *porcine endothelin*, or *porcine-ET*. Subsequent studies showed, however, that it is also produced by human aortic endothelial cells, so that for a time it became known as *porcine/human endothelin*. However, it is now clear that the amino acid sequence which characterizes this particular endothelin (Figure 2.2) is *not species specific*. Accordingly, the term endothelin-1 (or ET-1) has been introduced to denote the twenty-one amino acid configuration shown in Figure 2.2.

Endothelin-2

This was originally known as *human endothelin*, because its amino acid sequence was originally detected in the human genome. As with endothelin-1, however, the amino acid sequence of endothelin-2 (Figure 2.4) is not peculiar to a particular species: instead it defines the chemistry of endothelin-2. Hence this term (*endothelin-2*) should be used.

Endothelin-3

This was originally known as *rat endothelin*, because its amino acid sequence was originally detected in a rat genomic library. Once again, with the benefit of

Table 2.3. Amino acids present in the endothelins

Name	Abbreviation	
	Three letter code	Single letter code
Asparagine	Asn	N
Aspartic acid	Asp	D
Cysteine	Cys	C
Glutamic acid	Glu	E
Histidine	His	H
Isoleucine	Ile	I
Leucine	Leu	L
Lysine	Lys	K
Methionine	Met	M
Phenylalanine	Phe	P
Serine	Ser	S
Threonine	Thr	T
Tryptophan	Trp	W
Tyrosine	Tyr	Y
Valine	Val	V

hindsight, use of the term *endothelin-3* is more appropriate, since it is specific for the amino acid sequence shown in Figure 2.5, and is not species specific.

VIC [(Mouse) β Endothelin]

The continued use of the synonym VIC (vasointestinal contractor) is acceptable, because the predominant characteristic of this endothelin is its potency as a constricting agent for intestinal smooth muscle. It is not impossible, however, to imagine that it might ultimately become known as endothelin-4.

Hence throughout this book the terms endothelin-1, endothelin-2, endothelin-3 and VIC relate to the amino acid sequences of the various endothelin isoforms. These isoforms are *not* species specific, and they will be referred to as:
endothelin-1 (ET-1);
endothelin-2 (ET-2);
endothelin-3 (ET-3); and
VIC.

Molecular Manipulations Which Modify the Contractile Activity of Endothelin-1

Apart from the differences in contractile potency which result from the different amino acid compositions of the various endothelins (Figure 2.2, and 2.4 – 2.6) experiments with synthetic analogues of endothelin-1 have helped to shed some light on which regions of the molecule dominate its activity. For example, experiments along these lines have shown that:

(i) removal of the terminal tryptophan residue of endothelin-1 *reduces* its constrictor activity by a factor of approximately ten thousand;
(ii) removal of the terminal five amino acids results in a total loss of activity;
(iii) cleavage of the lysine residue at position *9* causes a three fold loss of activity;
(iv) alkylation or reduction of the disulphide bonds (between residues *1* and *15*, and *3* and *11*) causes a large (one thousand fold) reduction in activity, and
(v) the C-terminal hexapeptide (histidine, leucine, aspartic acid, isoleucine, isoleucine, tryptophan), although a highly conserved sequence in the endothelins, lacks any appreciable vasoconstrictor activity (Nakajima et al. 1989a, b; Kimura et al. 1989; Maggi et al. 1989b).

When consolidated, these findings indicate that not only the two pairs of intrachain cysteine linkages but also the C-terminal hexapeptide, including tryptophan, and the charged residues at positions 8 – 10 (Figure 2.2) are important determinants of the biological activities of these polypeptides.

The Relative Vasoconstrictor Activities of the Various Endothelins

Each of the endothelins is capable of causing vasoconstriction when administered in the appropriate dose. As far as their relative potencies are concerned ET-2 > ET-1 > VIC >> ET-3.

Summary

1. The endothelins are a family of acidic polypeptides (endothelin-1, endothelin-2, endothelin-3, and VIC). The isoforms are not species specific but are characterized by their particular amino acid sequence. They each contain:
   (a) twenty-one amino acids;
   (b) four cysteine residues, at the fixed postions of *1, 3, 11* and *15*;
   (c) two pairs of disulphide bonds between residues *1* and *15*, and *3* and *11* (Figure 2.2, and 2.3–2.6);
   (d) a cluster of charged residues at positions *8–10* (Figure 2.2, and 2.3–2.6); and
   (e) a terminal hexapeptide carboxy tail (C) with a terminal tryptophan residue.
2. Although each of these polypeptides produces a vasoconstrictor response, they differ in potency, ET-2 > ET-1 > VIC >> ET-3.
3. Manipulation of the amino acid composition and cleavage of the endothelin-1 molecule has shown that:
   (a) loss of the terminal tryptophan residue,
   (b) disruption of the disulphide bonds,
   (c) enzymatic cleavage of the molecule at position *9* (lysine), and
   (d) replacement of the charged residues at positions *8–10* reduces its potency as a vasoconstrictor.

In addition
   (e) the terminal hexapeptide assembly of amino acids (histidine, leucine, aspartic acid, isoleucine, isoleucine and tryptophan) of endothelin-1 lacks any appreciable constrictor activity.

Taken together, it seems that the vasoconstrictor activity of the endothelins reflects the combined effect of the presence of the intrachain linkages, the charged residues in the loop (positions *8–10*), and the presence of the C-terminal hexapeptide with its terminal tryptophan residue.

Chapter 3
The Biosynthesis of the Endothelins

"Logical consequences are the scare-crows of fools and the beacons of wise men".
T. H. HUXLEY, in: Science and Culture

No sooner had Yanagisawa and his colleagues established the amino acid sequence of endothelin-1, than they began investigating the biosynthetic pathways involved in its production. From the very start they assumed that since endothelial cells contain few secretory granules, the polypeptide must be produced within the endothelial cells by *de novo* synthesis. This is quite different from the situation which applies to the atrial natriuretic peptide (ANP) which, after being synthesized in the atrial cardiocytes, is stored there in granules for subsequent release (de Bold et al. 1981). A search of the cDNA library constructed for porcine aortic endothelial cells provided these investigators with the clue they needed – the presence of a mRNA-encoded nucleoside sequence containing the exact twenty-one amino acid sequence of the porcine endothelin molecule (Figure 1.1) they had

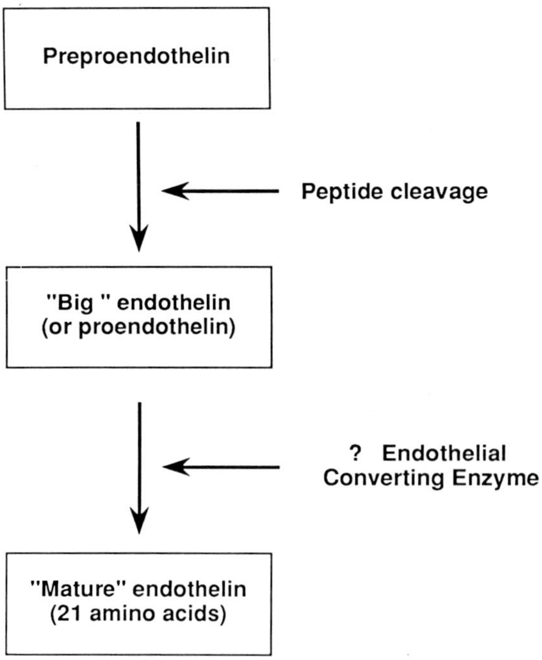

Fig. 3.1. Schematic representation of the biosynthesis of the endothelins (after Yanagisawa et al. 1988a)

isolated. Within a few weeks Yanagisawa and his colleagues had unravelled the steps involved in it's production.

The Biosynthesis of the Endothelins

The major steps involved in the synthesis of endothelin-1, -2, and -3, and VIC are summarized in Figure 3.1. They include:

(i) the production of the appropriately coded mRNA,
(ii) the conversion of a large mRNA-encoded precursor (*preproendothelin*) to a smaller unit – *"big" endothelin* (or proendothelin), followed by
(iii) the conversion of *"big"* endothelin to the *"mature"* twenty-one amino acid form described in the last chapter.

In many ways this method of production resembles that involved in the production of the atrial natriuretic peptide (ANP), where a pre-pro form of ANP, containing one hundred and fifty two amino acids, is first cleaved to release pro-ANP, from which the twenty eight amino acid peptide, ANP, is produced by cleavage of the carboxyterminal (Figure 3.2).

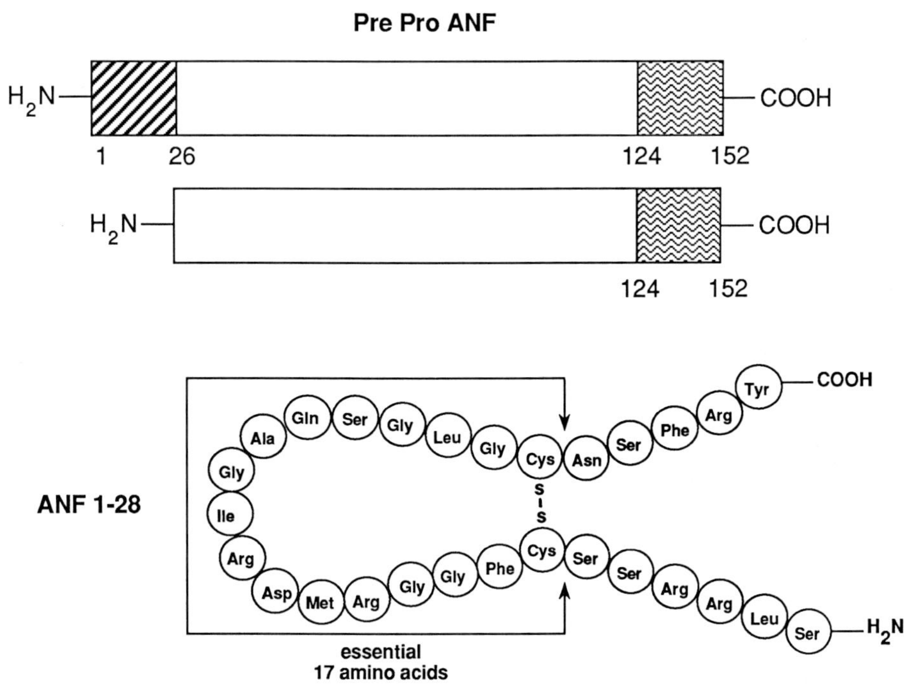

Fig. 3.2. Schematic representation of the events involved in the biosynthesis of the atrionatriuretic peptide (ANP – also known as ANF). Note the similarities between this scheme and that shown on Figure 3.1, for the endothelins (after Edwards et al. 1987)

Table 3.1. Production of endothelin-1

	Endothelin-1 (porcine) mRNA	*Endothelin-1* (human) mRNA
	↓	↓
Preproendothelin	203 amino acids	212 amino acids
	↓	↓
"Big" endothelin	39 amino acids	38 amino acids
	↓	↓
"Mature" endothelin	21 amino acids	21 amino acids

In both cases the liberation of "mature" endothelin from "big" endothelin involves the proteolytic cleavage of a tryptophan-valine link in "big" endothelin (see Table 3.2)

Endothelin-1 (Porcine and Human)

The chemical composition of "mature" porcine and human endothelin-1 are identical but there are some minor differences (Table 3.1) in the amino acid composition of their immediate precursors. The steps involved in their synthesis therefore need to be described separately, although the overall strategy is the same, the starting point in each case being the mRNA encoding of the mature endothelin sequence.

Porcine Endothelin-1

Porcine preproendothelin obtained from porcine aortic endothelial cells contains two hundred and three amino acids (Table 3.1). Residues *53* – *73* are occupied by the endothelin-1 sequence and are flanked (Figure 3.3) on one side (residues *51* and *52*) by a pair of basic amino acids – lysine at position *51* (Lys^{51}) and arginine at position *52* (Arg^{52}). Residue *53* is occupied by the first cysteine residue of the endothelin sequence. At the distal end of the endothelin-1 sequence, and hence adjacent to the tryptophan residue (Trp^{73}) which will ultimately terminate the endothelin-1 sequence, position *74* of preproendothelin is occupied by a valine residue (Val^{74}, Figure 3.3). Further along, at residues *92* and *93* is another pair of basic amino acids. In this case both loci are occupied by arginine residues (Figure 3.3).

Converting preproendothelin to "big" endothelin (or "proendothelin" as it is sometimes called), involves the enzymatic cleavage of the two pairs of basic residues at positions *51* and *52*, and *92* and *93* of the preproendothelin complex. Presumably cleavage involves the action of an endopeptidase. Such a step is common to the production of many peptides and neuropeptides, but in this instance the end-product is the immediate precursor of "mature" endothelin-1. Yanagisawa and his colleagues called this precursor *"big endothelin"*, but the term *proendothelin* is equally applicable.

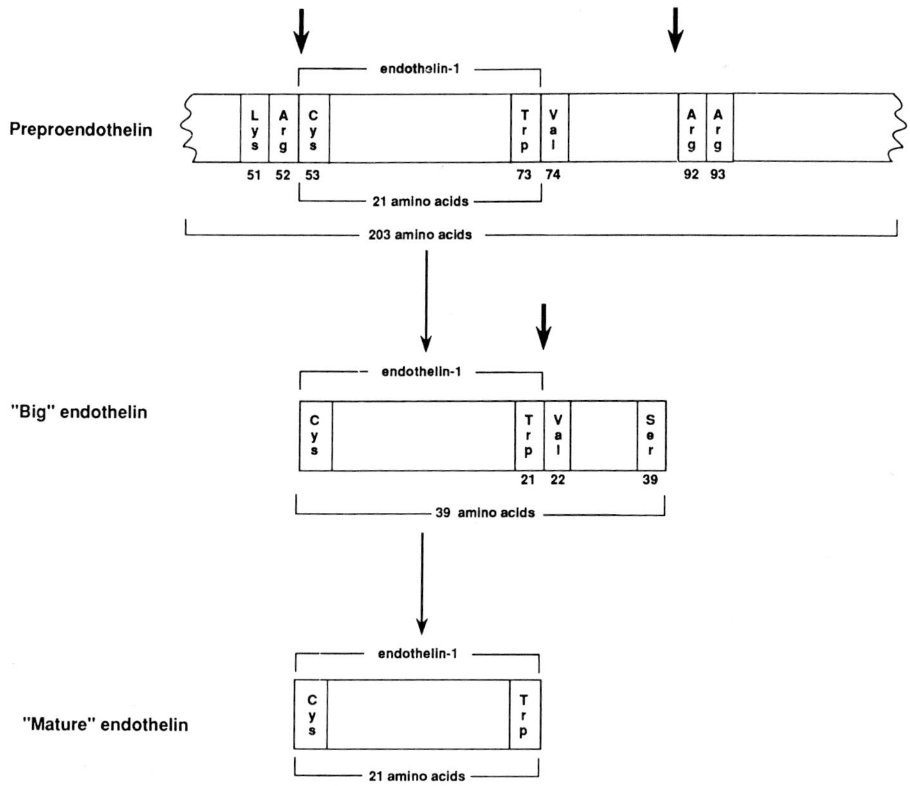

Fig. 3.3. Sites of proteolytic cleavage involved in the biosynthesis of mature endothelin from the porcine precursors. The heavy arrows mark the sites of enzymatic cleavage. Lys (lysine); Arg (arginine); Cys (cysteine); Trp (tryptophan); Val (valine); Ser (serine)

Biosynthesis of Porcine Endothelin-1

Porcine "big" endothelin-1 contains thirty-nine amino acids (Table 3.1) and has a molecular weight of 22,478 daltons. It's thirty-nine amino acids are arranged as follows:

(i) residues *1–21*: endothelin-1 sequence, with tryptophan at residue *21*
(ii) residue *22*: valine
(iii) residues *23–29*: other non-essential amino acids

The conversion of this *"big"* endothelin-1 to the *"mature"* endothelin-1 involves (Figure 3.3) an unusual proteolytic cleavage between the terminal tryptophan residue of the endothelin sequence (Trp^{21}) and the valine residue at position *22* (Val^{22}). Investigators now believe that they have identified the enzyme which is responsible for this cleavage (Sawamura et al. 1989) but as yet it has not been fully identified. It may be an endopeptidase with chymotrypsin-like specificity, although there is also reason to believe that it could be a pepsin-like aspartic protease (Takaoka et al. 1990). At the moment it is known simply as *"the endothelin-converting enzyme"* (Yanagisawa et al. 1988a).

There are at least three reasons for believing that the scheme outlined in Figure 3.1 might correctly describe the sequence of events involved in the synthesis of endothelin-1. These reasons are:

(i) the existence of the endothelin-1 sequence in the mRNA coding of aortic endothelial cells (Yanagisawa et al. 1988a),

(ii) the detection of "big" endothelin (39 amino acids), and endothelin-1 (21 amino acids), along with the truncated residues of "big" endothelin (residues *22–39*) in the culture supernatant of porcine, bovine and human endothelial cells (Sawamura et al. 1989; Emori et al. 1989a, b), and

(iii) the ability of synthetic chymotrypsin to release endothelin-1 from "big" endothelin (McMahon et al. 1989).

Human Endothelin-1

Location of Gene Encoding Endothelin-1 mRNA on Human Chromosome 6

The gene encoding for human endothelin-1 occurs on chromosome *6* (Bloch et al. 1989b). This gene must be widely expressed, because endothelin-1 mRNA has been identified in a variety of tissues, including umbilical vein endothelial cells (Inoue et al. 1989a, b), foetal lung, foetal spleen, foetal pancreas and foetal heart (atria and ventricles) (Bloch et al. 1989a, b). The "mature" human endothelin-1 sequence has also been located in the cDNA library of the human placenta (Itoh et al. 1988).

Biosynthesis of Human Endothelin-1

The general scheme for the production of human endothelin-1 parallels that of porcine endothelin-1 (Table 3.1) and the end-product is identical. There are, however some minor differences in the biosynthetic pathway (Table 3.2). Thus:

(i) whereas porcine preproendothelin contains 203 amino acids, human preproendothelin is slightly larger; it contains (Table 3.1) 212 residues (Itoh et al. 1988);

(ii) "big" porcine endothelin-1 contains 39 amino acids whereas "big" human endothelin-1 contains only 38 (Tables 3.1, 3.2); and

(iii) in porcine "big" endothelin-1 residue *28* is occupied by isoleucine (Ile), whereas in human "big" endothelin-1 residue *28* is occupied by valine (Table 3.2).

Despite these minor differences the end-product is the same, and in each case the liberation of the "mature" endothelin-1 from the immediate precursor involves proteolytic cleavage between the terminal tryptophan residue of the "mature" endothelin-1 sequence and the adjacent valine residue of the "big" endothelin-1 molecule (Trp^{21} Val^{22} of "big" endothelin (Table 3.2)).

Table 3.2. Amino acid sequence of porcine and human "big" endothelin

Residue	Porcine	Human
1	Cysteine	Cysteine
2	Serine	Serine
3	Cysteine	Cysteine
4	Serine	Serine
5	Serine	Serine
6	Leucine	Leucine
7	Methionine	Methionine
8	Aspartic acid	Aspartic acid
9	Lysine	Lysine
10	Glutamic acid	Glutamic acid
11	Cysteine	Cysteine
12	Valine	Valine
13	Tyrosine	Tyrosine
14	Phenylalanine	Phenylalanine
15	Cysteine	Cysteine
16	Histidine	Histidine
17	Leucine	Leucine
18	Aspartic acid	Aspartic acid
19	Isoleucine	Isoleucine
20	Isoleucine	Isoleucine
21	Tryptophan	Tryptophan
22	Valine	Valine
23	Asparagine	Asparagine
24	Threonine	Threonine
25	Proline	Proline
26	Glutamic acid	Glutamic acid
27	Histidine	Histidine
28	Isoleucine*	Valine*
29	Valine	Valine
30	Proline	Proline
31	Tyrosine	Tyrosine
32	Glycine	Glycine
33	Leucine	Leucine
34	Glycine	Glycine
35	Serine	Serine
36	Proline	Proline
37	Serine*	Arginine*
38	Arginine	Serine
39	Serine	

Note that the proteolytic cleavage involves disruption of the tryptophan-valine sequence in both cases. Human "big" endothelin-1 differes from porcine "big" endothelin-1 in containing valine instead of isoleucine at residue *28* (*) and arginine instead of serine at position *37*. Thus residue serine^{37} in porcine "big" endothelin-1 is missing from human "big" endothelin-1

Endothelin-2 and Endothelin-3

mRNA coding for these two endothelins, as well as for endothelin-1, has been detected in human, porcine and rat genomes. Cloning and sequence analysis has shown that one mRNA sequence codes for endothelin-1, another for endothelin-2 and another for endothelin-3, irrespective of species. Hence the term *rat endothelin*, when used as a pseudonym for endothelin-3, is really a misnomer, since "rat" endothelin is a type of endothelin (endothelin-3) and is *not species specific* (Yanagisawa and Masaki 1989a, b). However, as already mentioned in Chapter 2, vascular endothelial cells, although capable of expressing endothelin-1, do *not* express the other endothelins – endothelin-2, -3 and VIC.

Location of Gene Encoding Endothelin-3 – mRNA on the Human Chromosome 20

The gene encoding for endothelin-3 mRNA is located on human chromosome number *20* (Bloch et al. 1989a) – whereas the gene encoding for endothelin-1 mRNA occurs in chromosome number *6* (Bloch et al. 1989b). This difference must mean that the two genes are not genetically linked, which is surprising because endothelin-3 shares many of the properties of endothelin-1 – including an ability to induce mitogenesis and to cause vaso- and bronchoconstriction. The chromosomal location of the gene encoding for endothelin-3 has yet to be identified.

The tissue distribution of endothelin-3 mRNA differs from that of endothelin-1, although, as Table 3.3 shows, there is some overlap. Thus endothelin-3 mRNA appears to be absent from umbilical vein endothelial cells (Bloch et al. 1989a, b) whereas the mRNA for endothelin-1 is present (Inoue et al. 1989b). However foetal lung, spleen and pancreas all contain relatively high concentrations of mRNA for both endothelin-1 and -3 (Bloch et al. 1989a, b). The hypothalamus and kidneys are other tissues in which mRNA for endothelin-3, but not endothelin-1, is readily detectable, even though the levels are quite low. The mere fact of mRNA encoding for endothelin-3 being present in the hypothalamus is of considerable interest,

Table 3.3. Tissues in which mRNA for endothelin-1 and endothelin-3 has been identified

Tissue distribution	
mRNA for endothelin-1	mRNA for endothelin-3
spleen	spleen
lung	lung
pancreas	pancreas
heart (atria and ventricles)	hypothalamus
endothelium of umbilical vein	kidney
human amniotic cells	

(Data from Bloch et al. 1989a, b. The tissue used for this study was predominately foetal tissue). The data for human amniotic cells is from Sunnergren et al. 1990. This list is not meant to be exhaustive. It is included simply to show that the tissue distribution of the mRNA's for these two isoforms of endothelin is not identical, although in the case of the spleen, lungs and kidneys there is some overlap because it re-enforces the hypothesis of this particular endothelin being involved in neurotransmission, or neurosecretion.

Biosynthesis of Endothelin-3

The sequence of events involved in the synthesis of endothelin-2 and endothelin-3 parallels that involved in the production of endothelin-1. The starting point is again the mRNA encoded preproendothelin (Yanagisawa et al. 1988b; Ohkubo et al. 1990) which, by proteolytic cleavage of basic pairs, is converted into the smaller and intermediate "big" endothelin. This process is followed by another proteolytic cleavage to release the "mature" endothelin. The sequence has been worked out in detail for endothelin-3 but not, as yet for endothelin-2. In the case of endothelin-3 there are some minor differences relative to the system developed for endothelin-1.

The differences include:
(i) whereas the basic pairs of amino acids which must be removed to liberate "big" porcine or human endothelin-1 from their relative precursors are lysine and arginine at one end, and a pair of arginine residues at the other, for endothelin-3, the lysine-arginine of the endothelin-1 precursor is replaced (Table 3.4) by arginine-arginine, and
(ii) the liberation of "mature" endothelin-3 from "big" endothelin-3 involves the proteolytic cleavage of a tryptophan-isoleucine sequence (compared with the tryptophan-valine sequence of "big" endothelin-1, Table 3.4).

Vasoactive Intestinal Contractor Endothelin (VIC)

This endothelin-like polypeptide differs from endothelin-1 in only three amino acids. Thus, at positions *9, 11* and *12* in the mature form, the serine, leucine and methionine residues of endothelin-1 become asparagine, tryptophan and leucine. The four cysteine residues and the terminal tryptophan are retained (Figure 2.5 and Table 3.4).

The Biosynthesis of VIC

The mRNA encoding for VIC has been located (Saida et al. 1989) in the mouse genome and its site of expression identified as the intestine. In the case of VIC:
(i) the dibasic pair immediately adjacent to the first cysteine residue of the mature VIC sequence in the prepropr ecursor is probably a pair of arginine residues – as happens with (rat endothelin) prepropr ecursor (Table 3.4). Also
(ii) the amino acid adjacent to the terminal tryptophan residue of "mature" VIC is valine, resembling the situation described for endothelin-1 and differing from that of endothelin-3 (Table 3.4).

Table 3.4. Comparison of the primary structures of the endothelins and their immediately adjacent amino acids in the preproendothelin precursor

	Porcine ET-1	Human ET-1	ET-3 (rat)	VIC
	Arg	Arg	His	Leu
	Arg	Arg	Arg	Arg
	Ser	Ser	Pro	Phe
	Lys	Lys	Arg	Arg
	Arg	Arg	Arg	Arg
	Cys	Cys	Cys	Cys
	Ser	Ser	*Thr*	Ser
	Cys	Cys	Cys	Cys
	Ser	Ser	*Phe*	*Asn*
	Ser	Ser	*Thr*	Ser
	Leu	Leu	*Tyr*	*Trp*
	Met	Met	*Lys*	*Leu*
	Asp	Asp	Asp	Asp
	Lys	Lys	Lys	Lys
	Glu	Glu	Glu	Glu
	Cys	Cys	Cys	Cys
	Val	Val	Val	Val
	Tyr	Tyr	Tyr	Tyr
	Phe	Phe	*Tyr*	Phe
	Cys	Cys	Cys	Cys
	His	His	His	His
	Leu	Leu	Leu	Leu
	Asp	Asp	Asp	Asp
	Ile	Ile	Ile	Ile
	Ile	Ile	Ile	Ile
21	Trp	Trp	Trp	Trp
	Val	Val	Ile	Val
	Asn	Asn	Asn	Asn

The boxes contain the twenty-one amino acid sequences of porcine and human endothelin-1 (ET-1), endothelin-3 (ET-3, rat) and vasointestinal constrictor (VIC – mouse β endothelin). The residues beyond the boxes relate to the amino acids adjacent to the endothelin sequences in the relevant preproendothelin molecule. Proteolytic cleavage occurs in two stages:

(i) removal of the dibasic pairs (lysine-arginine for porcine and human endothelin-1, and arginine-arginine for endothelin-3, and VIC); and

(ii) disruption of the trytophan-valine link for porcine and human endothelin-1 and VIC, compared with the disruption of a tryptophan-isoleucine link for endothelin-3

The Biological Significance of the Conversion of the "big" Endothelins to "Mature" Endothelins

The importance of the conversion of "big" endothelin to "mature" endothelin (Figure 3.1) can be illustrated simply by comparing the potency of "big" and "mature" endothelin-1 as coronary vasoconstrictor agents. "Big" endothelin-1 is about 140 times *less* potent than "mature" endothelin-1 (Kimura et al. 1989). Presumably this same situation applies to the other endothelins.

Summary

1. The mRNA encoding for the endothelins contains the amino acid sequence for the preproprecursors. Production of the active "mature" form of the endothelins (including VIC) involves liberating a smaller precursor form ("big" endothelin) by a process of proteolytic cleavage. This is then followed by another cleavage to liberate the appropriate "mature" twenty-one amino acid sequence. This twenty-one amino acid sequence can be identified in the mRNA encoded preproprecursor. The liberation of the "mature" endothelin can be likened, therefore, to the peeling away or removal of superfluous but adjacent amino acids.
2. The gene encoding human endothelin-1 mRNA occurs on human chromosome *6*, whereas the gene encoding endothelin-3 mRNA is located on chromosome *20*. This means that the two genes are not genetically linked.
3. The tissue distribution of mRNA's encoding for endothelin-1 and endothelin-3 are not identical, although there is some overlap.

Chapter 4

Other Chemically-Related Toxins

"If it were played upon a stage, now, I could condemn it as improbable fiction".
WILLIAM SHAKESPEARE, Twelfth Night (Scene III)

As soon as the essential chemical features of the endothelins (twenty-one amino acids, two pairs of intrachain disulphide bridges, four cysteine residues at fixed positions (*1, 3, 11* and *15*), a hydrophobic tail and a terminal tryptophan residue) had been established it became apparent that their chemical profile is not unique (Lee and Chiappinelli 1988). To the contrary, it bears a remarkable and almost uncanny resemblance to that of the toxic peptides which Kochva and his colleagues (Kochva et al. 1982) and Bdolah et al. (1989) have isolated from the venom of the burrowing asp – a snake of the genus *Atractaspis engaddensis*. These toxic peptides are now known as *sarafotoxins* – after the common Hebrew name for the snake. Probably they deserve a place in the history books, because it was the venom of the burrowing asp which caused the premature, but self inflicted death of Cleopatra. Some idea of their potency can be gauged by the fact that ten nanograms per gram body weight is sufficient to cause sustained coronary vasospasm, cardiac arrest and death.

The sarafotoxins are not the only toxins which chemically resemble the endothelins. In many ways the polypeptide *Apamin* which occurs in bee venom resembles them, although (Table 4.1) it only contains eighteen, and not twenty-one, amino acids (Lazdunski 1983).

Table 4.1. Naturally-occuring peptide based toxins which modulate Ca^{2+} availability

Toxin	Source	Chemistry	Site of action
Apamin	Bee venom	Polypeptide (18 amino acids)	L-type Ca^{2+} channel Ca^{2+}-activated (K^+ channel)
Sarafotoxins (a–d)	*Atractaspis engadd.* (burrowing asp)	Polypeptide (21 amino acids)	Endothelin/sarafotoxin receptor
Atrotoxin	*Crotalus atrox.* (rattlesnake)	Polypeptide (M. W. > 15,000)	L-type Ca^{2+} channel

For differences between L-, T- and N-type Ca^{2+} channels see references Miller (1987) and Nayler (1988)

Table 4.2. Amino acid sequences of the endothelins and the sarafotoxins

	ET-1	ET-2	ET-3	SRTX-a	SRTX-b	SRTX-c	SRTX-d	VIC
1	Cys	Cys	Cys	Cys	Cys	Cys	Cys	Cys
	Ser	Ser	Thr	Ser	Ser	Thr	Thr	Ser
	Cys	Cys	Cys	Cys	Cys	Cys	Cys	Cys
	Ser	Ser	Phe	Lys	Lys	Asn	Lys	Asn
5	Ser	Ser	Thr	Asp	Asp	Asp	Asp	Cer
	Leu	Trp	Tyr	Met	Met	Met	Met	Trp
	Met	Leu	Lys	Thr	Thr	Thr	Thr	Leu
	Asp	Asp	Asp	Asp	Asp	Asp	Asp	Asp
	Lys	Lys	Lys	Lys	Lys	Glu	Lys	Lys
10	Glu	Glu	Glu	Glu	Glu	Glu	Glu	Glu
	Cys	Cys	Cys	Cys	Cys	Cys	Cys	Cys
	Val	Val	Val	Leu	Leu	Leu	Leu	Val
	Tyr	Tyr	Tyr	Asn	Tyr	Asn	Tyr	Tyr
	Phe	Phe	Tyr	Phe	Phe	Phe	Phe	Phe
15	Cys	Cys	Cys	Cys	Cys	Cys	Cys	Cys
	His	His	His	His	His	His	His	His
	Leu	Leu	Leu	Gln	Gln	Gln	Gln	Leu
	Asp	Asp	Asp	Asp	Asp	Asp	Asp	Asp
	Ile	Ile	Ile	Val	Val	Val	Ile	Ile
20	Ile	Ile	Ile	Ile	Ile	Ile	Ile	Ile
	Trp	Trp	Trp	Trp	Trp	Trp	Trp	Trp

Amino acid sequence of SRTXa–d compared with that of the other sarafotoxins and the endothelins (from Bdolah et al. 1989)

The Sarafotoxins

The Chemistry of the Sarafotoxins

As with the endothelins, the sarafotoxins are a *family* of polypeptides. Four isoforms have been identified so far. They are:
(i) sarafotoxin S6*a* (or SRTX-a);
(ii) sarafotoxin S6*b* (or SRTX-b);
(iii) sarafotoxin S6*c* (or SRTX-c); and
(iv) sarafotoxin S6*d* (or SRTX-d).

Each contains twenty-one amino acids, two pairs of intrachain disulphide bridges, a hydrophobic tail, and a C-terminal tryptophan residue. Their chemical homology with the endothelins extends beyond this, however, since they contain many common amino acids (Table 4.2). This similarity probably means that the endothelins and the sarafotoxins share a common ancestry. However, whilst it is easy to imagine that the presence of the sarafotoxins in the venom of the burrowing asp provides it with an effective means of dealing with its predators, the evolutionary significance of the endothelins is less certain – unless it provided a means of controlling injury-induced haemorrhage.

The toxic components of many other naturally occurring venoms are polypeptides – including some snake venoms and the neurotoxin of the scorpions. However, although these polypeptides generally resemble the endothelins and sarafotoxins in containing four disulphide bridges they are much larger molecules. Scorpion neurotoxins, for example, contain more than sixty polypeptides (sixty-three to sixty-four, to be precise). Their major site of action is also quite different from that of the endothelins and sarafotoxins, since they act primarily on the excitable membranes of axons rather than the cardiovascular system and often interact directly with the sodium-conducting channels

Sarafotoxin S6a (or SRTX-a)

This is highly lethal. Death is by cardiac arrest (Weiser et al. 1984; Kloog et al. 1988). In common with the other sarafotoxins and endothelins (Table 4.2, and Figure 4.1) it contains cysteine residues at positions *1, 3, 11* and *15*, and a tryptophan residue at position *21*. There are, however, eight amino acid substitutions relative to endothelin-1, at positions *4, 5, 6, 7, 12, 13, 17* and *19*. Thus sarafotoxin S6a could be described as: Lys^4, Asp^5, Thr^6, Ser^7, Leu^{12}, Asn^{13}, Gln^{17}, Val^{19} ET-1, where Lys = lysine, Asp = aspartic acid, Thr = threonine, Ser = serine, Leu = leucine, Gln = glutamine and Asn = asparagine. Despite these substitutions, however, the cluster of negatively and positively charged residues at positions *8–10* (Table 4.2 and Figure 4.1) are retained.

Sarafotoxin S6b (or SRTX-b)

This differs from sarafotoxin S6b in only one amino acid substitution – the presence of a tyrosine instead of an asparagine residue at position *13*. Thus sarafotoxin S6b is really Tyr^{13} sarafotoxin S6a, or Lys^4, Asp^5, Thr^6, Ser^7, Leu^{12}, *Tyr^{13}*, Gln^{17}, Val^{19} ET-1 (with Tyr = tyrosine) (Table 4.2, and Figure 4.1). The charged residues at positions *8–10* are retained, as are the cysteine residues at positions *1, 3, 11* and *15*, and the terminal tryptophan. In all, there have been seven amino acid substitutions relative to endothelin-1.

Sarafotoxin S6c (or SRTX-c)

Superficially this isoform bears a close resemblance to endothelin-1 and the other sarafotoxins. However (Figure 4.1) it differs from the others in that the amino acid sequence involving residues *8–10* is different, with the replacement of the lysine residue at position *9* in endothelin-1, sarafotoxin S6a and S6b, by a glutamic acid residue (Glu^9). There are nine other amino acid substitutions relative to endothelin-1 (Thr^2, Asn^4, Asp^5, Met^6, Thr^7, Glu^9, Leu^{12}, Asn^{13}, Gln^{17}, Val^{19}), relative to

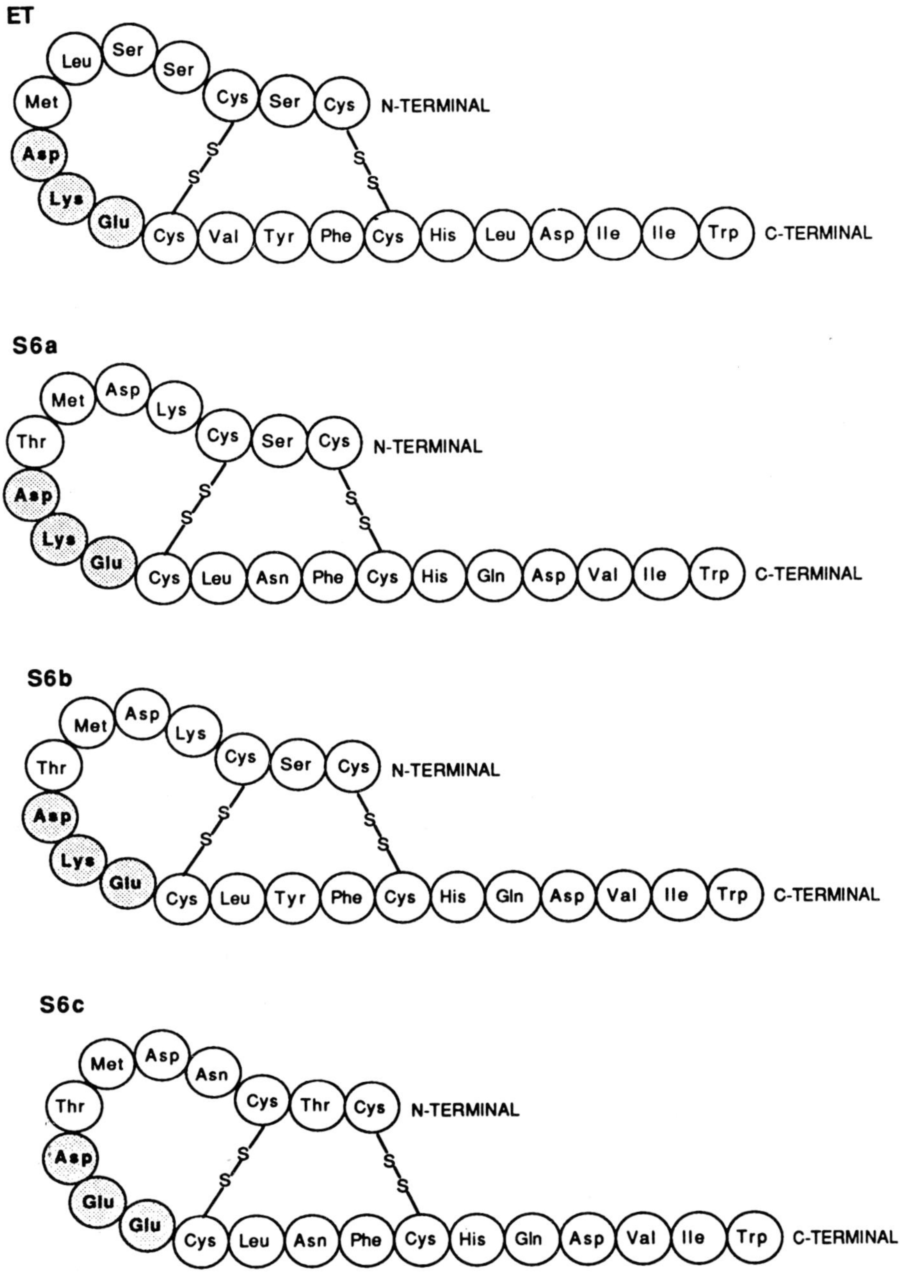

Fig. 4.1. Chemical structures of endothelin-1 (ET) and sarafotoxin S6a, b, and c (SRTX-a, -b and -c). Note the presence of twenty-one amino acids, two intrachain disulphide bonds, cysteine residues at positions *1, 3, 11* and *15*, charged residues at positions *8–10*, the hydrophobic tail and a terminal tryptophan residue

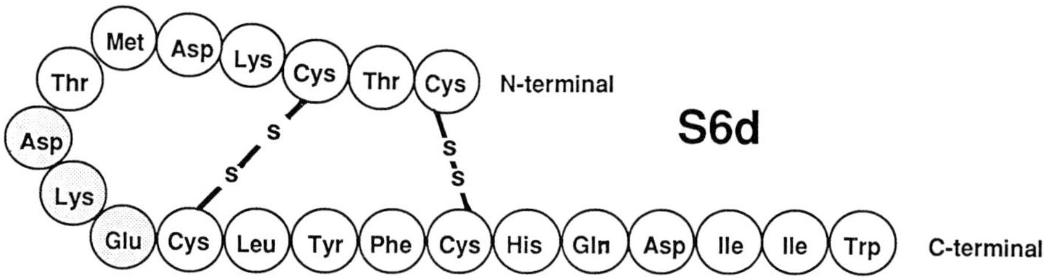

Fig. 4.2. Chemical structure of sarafotoxin S6d (SRTX-d). Note the presence of amino acid substitutions at postions *2, 4, 5, 6, 7, 12* and *17*, relative to endothelin-1. See Figure 4.1 for endothelin-1 structure

endothelin-1 (Thr = threonine, Asn = asparagine, Asp = aspartic acid, Met = methionine, Glu = glutamic acid, Leu = leucine, Gln = glutamine, Val = valine). Thus sarafotoxin S6c becomes Thr^2, Asn^4, Asp^5, Met^6, Thr^7, Glu^9, Leu^{12}, Asn^{13}, Gln^{17}, Val^{19} – ET-1 (Figure 4.1). It is perhaps noteworthy that sarafotoxin S6c lacks lysine and arginine residues (Figure 4.1), and because of this it is more acidic than the other sarafotoxins, and endothelin-1.

Sarafotoxin S6d (or SRTX-d)

This sarafotoxin is the most recently discovered member of the group (Bdolah et al. 1989) and from a purely academic point of view it is important because it sheds light on which of the amino acids are of importance as far as the biological activity of these polypeptides is concerned.

Sarafotoxin S6d differs from endothelin-1 (Table 4.2 and Figure 4.2) in only seven of its amino acids – at positions *2, 4, 5, 6, 7, 12* and *17*. Hence it can be described as: Thr^2, Lys^4, Asp^5, Met^6, Thr^7, Leu^{12}, Gln^{17} ET-1.

When compared with sarafotoxin S6a, the S6d isoform differs in only three amino acids, isoleucine (Ile) instead of valine (Val) at position *19*; tyrosine (Tyr) instead of asparagine (Asn) at position *13*, and threonine (Thr) instead of serine (Ser) at position *2*. This substitution of threonine for serine at position *2* may be biologically important because on the three occasions in which it occurs (S6c, S6d and ET-3, Table 4.2) the polypeptides are relatively weak members of the group. Thus, for example (Table 4.3) the LD_{50} for S6d (dose required to produce death in fifty percent of the test animals) in mice is at least thirty times greater than that of the other sarafotoxins, and of endothelin-1.

Mode of Action

Radioligand binding studies have shown that:

(a) The sarafotoxins interact with membrane located receptors (Kloog et al. 1988; Ambar et al, 1989a, b);

Table 4.3. LD_{50} values for endothelin-1 and the sarafotoxins

	Polypeptide				
	ET-1	S6a	S6b	S6c	S6d
LD_{50}* (ng/g bodyweight)	10	10	10	350	350

* LD_{50} is the concentration required to produce death in fifty percent of the animals which, in this case, were mice. Data from Bdolah et al. (1989)

S6a = SRTX-a, S6b = SRTX-b, S6c = SRTX-c, S6d = SRTX-d

(b) these receptors are also capable of binding endothelin-1 (Gu et al. 1989a; Ambar et al, 1989a, b);
(c) there are approximately the same number of receptors for sarafotoxin S6b as there are for endothelin-1;
(d) these receptors do not bind other biologically active peptides; and
(e) as with the endothelins (Chapter 6) the interaction of the sarafotoxins results in activation of the inositol triphosphate pathway.

Clearly, the similarities between the endothelins and the sarafotoxins cannot be ignored, the more so because they must be considered against the background of their remarkably similar chemical profiles. Why, then, in man are the sarafotoxins far more toxic than the endogenous endothelins? There are at least two answers to this question:
(i) the small differences in amino acid sequences may account for the higher toxicity of the sarafotoxins; or
(ii) the tissue content of the endothelins may be tightly regulated. This would not be impossible, because the endothelins have a relatively short half-life in blood plasma, and the half life of the mRNA which is encoded for them is also relatively short (about fifteen minutes). Probably the endothelins act as local autocoids, being produced close to their site of utilization, and only when needed,
(iii) as far as the vaso- and broncho-constrictor effect of the endothelins are concerned the response is often blunted by the endothelin-induced release of endogenous vasodilator agents, including prostacyclin, the endothelial-derived relaxing factor and the atrial natriuretic peptide.

Apamin

This is a neurotoxin present in bee venom. Superficially it resembles the endothelins in possessing two pairs of intrachain disulphide bridges, and cysteine residues at positions *1, 3, 11* and *15* (Figure 4.3). However it differs from the endothelins (and the sarafotoxins) in that:

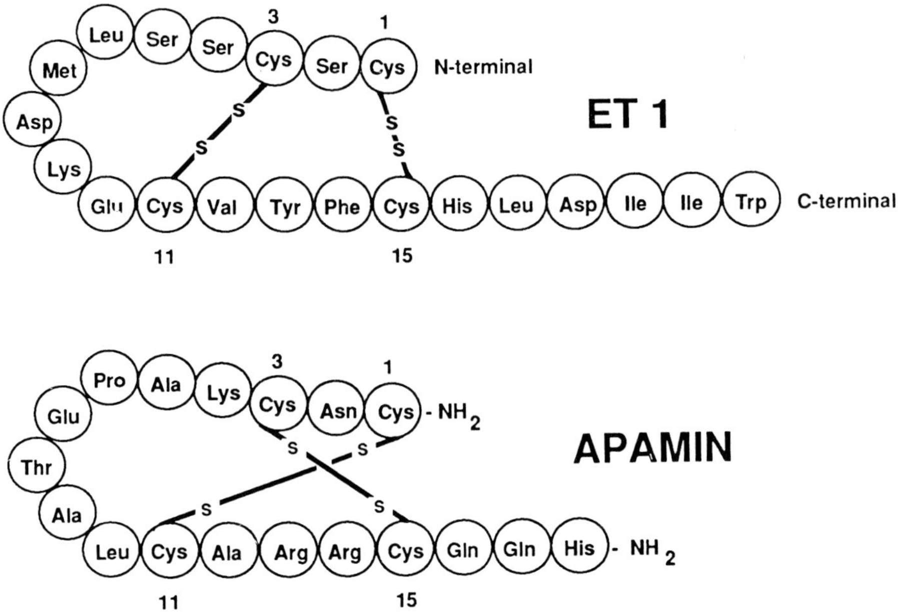

Fig. 4.3. Comparison between the chemical structure of endothelin-1, and that of the bee apamin toxin. Note that apamin contains only eighteen amino acids, that it lacks a terminal tryptophan residue and that the disulphide bonds link the cysteine residues at *1* and *11*, and *3* and *15*. The apamin molecule is actually arranged with an α helical core and regions of β-type turns (Lazdunski 1983)

(i) it contains eighteen (instead of twenty-one) amino acids;
(ii) the intrachain disulphide bridges are between residues *1* and *11*, and *3* and *15* (compared with bridges between residues *1* and *15*, and *3* and *11* in the endothelins and sarafotoxins);
(iii) it lacks a terminal tryptophan residue;
(iv) it does not interact with the endothelin-sarafotoxin sensitive receptors; and
(v) in marked contrast to the endothelins and the sarafotoxins it interacts directly with the slow calcium channels (Bkaily et al. 1985). In addition,
(vi) it has a direct inhibitory effect on the Ca^{2+}-dependent K^+ channels (Lazdunski 1983), a property which the endothelins and sarafotoxins lack, and
(vii) arginine residues at positions *13* and *14* (Figure 4.3) are essential for its biological activity.

Hence, although apamin is a potent polypeptide-based toxin, it does not really resemble either the sarafotoxins or the endothelins in its chemistry and mode of action.

Table 4.4. Molecular weights of some polypeptide-based toxins

Toxin	Source	M.W. (daltons)
Endothelin-1	Endothelial cells	2,492
Sarafotoxin S6b	Asp (snake)	2,562
Apamin	Bee venom	2,027
Atrotoxin	Rattle snake	> 15,000

M.W. = molecular weight; S6b = SRTX-b

Atrotoxin

This toxin is present in the venom of the rattle snake. Although a peptide, it does not belong to the sarafotoxin-endothelin family because:

(i) it is a relatively large polypeptide (M.W. 15,000 daltons) compared with 2491.8 for endothelin-1 (Table 4.4);
(ii) it has a strong basic charge (whereas the endothelins are acidic); and
(iii) it interacts *directly* with the Ca^{2+} channels.

Summary

As far as naturally occuring toxins are concerned:

(i) the sarafotoxins are chemically homologous with the endothelins. Both contain:
  (a) twenty-one amino acids;
  (b) two pairs of intrachain disulphide bonds;
  (c) cysteine residues at positions *1, 3, 11* and *15*;
  (d) a terminal tryptophan residue, and
  (e) charged residues at positions *8–10*.
(ii) both families of polypeptides interact with a commmon receptor, independent of the Ca^{2+} channels (Chapter 7);
(iii) in each case occupancy of the receptor results in activation of the phosphoinositol pathway (Chapter 6); and
(iv) in each case the physiological response is the same. It includes potent vasoconstriction and cardiotoxicity (Chapter 5).

Chapter 5
The Contractile Effect of Endothelin-1

"There must be a beginning of any great matter but the continuing until the end, until it be thoroughly finished, yields the glory."
From SIR FRANCIS DRAKE's dispatch to SIR FRANCIS WALSINGHAM, in 1575

Now that the chemical structure and the biochemical pathways involved in the biosynthesis of the endothelins have been described (Chapters 2 and 3) the next problem to address is their mode of action, because when injected into the circulation endothelin-1 causes a marked and unusually sustained pressor response (Yanagisawa et al. 1988a), accompanied by sustained coronary artery constriction (Figure 5.1, and Kurihara et al. 1989a, b; Clozel and Clozel 1989) and a significant reduction in renal blood flow (King et al. 1989; Lopez-Farre et al. 1989).

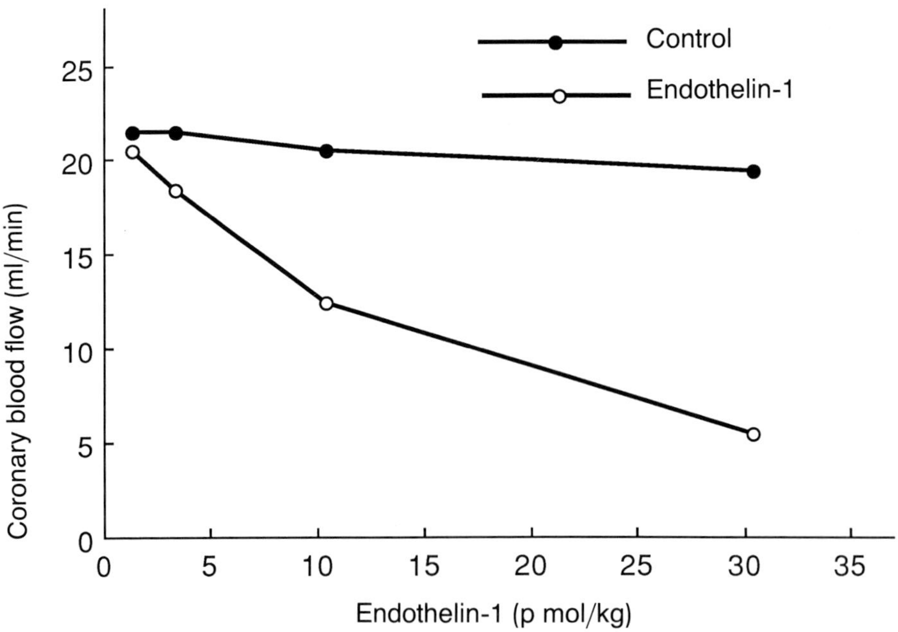

Fig. 5.1. Graph showing the effect of intracoronary injections of endothelin-1 on coronary blood flow in open-chest dogs. The doses of endothelin-1 were cumulative

Table 5.1. Effect of endothelin-1 on cytosolic Ca^{2+} in vascular smooth and cardiac muscle cells

Preparation	Response	Reference
Cultured rat aortic smooth muscle cells (A7-5 cell line)	Increase	Van Renterghem et al. 1988
Rat aorta	Increase	Sakata et al. 1989
Cultured rat aortic smooth muscle cells	Increase	Wallnofer et al. 1989
Rat cardiocytes	Increase	Hirata et al. 1989e

In these studies Ca^{2+}-sensitive dyes (Indo-1 or Fura-2) were used to measure cytosolic Ca^{2+}

Since the constrictor activity of this polypeptide is not mediated by alpha adrenergic, histaminergic, cholinergic or serotonergic mechanisms, and neither vasopressin nor angiotensin seem to be involved (Yanagisawa et al. 1988a) some other explanation must be sought to account for it. At the same time, an explanation must be provided for the fact that endothelin-1 exerts a positive inotropic effect on the heart (Ishikawa et al. 1988a). Both problems have a common answer: the endothelins, and particularly endothelin-1, *increase cytosolic* Ca^{2+}, and since tension development in cardiac and smooth muscle cells is Ca^{2+}-dependent, it is reasonable to assume that the increase in cytosolic Ca^{2+} is responsible for the enhanced contractile response.

Endothelin-1 is the endothelin which has been most widely used to investigate why these polypeptides increase cytosolic Ca^{2+} in smooth as well as cardiac muscle cells (Table 5.1). Of necessity, therefore, this and the next chapter will concentrate on this particular isoform. The present chapter is only intended to provide evidence of the constrictor effect of endothelin-1 on the vasculature and its positive inotropic effect on the heart. The mechanisms which may be involved are outlined in the next chapter.

Endothelin-1: Evidence of its Contractile Effect on Smooth Muscle

Vascular Smooth Muscle

Endothelin-1 exerts a potent, dose-dependent constrictor effect on vascular smooth muscle (Table 5.2). The effect is not species specific, and occurs under *in vivo* as well as *in vitro* conditions. Arteries, arterioles, veins and venules are all affected (Table 5.2), with veins being more sensitive than arteries (Table 5.3). Venules and arterioles, however, appear to be equisensitive (Fortes et al. 1989, and Table 5.4). Often the constrictor response is preceded by a transitory relaxation – caused, it is generally believed, by the release of the endothelial-derived relaxing factors, including EDRF and prostacyclin (Chapter 1). This dilator response is particularly apparent when relatively high concentrations of endothelin-1 are used (Cocks et al. 1989). As far as the coronary constrictor effect of endothelin-1 is concerned it is not species specific, having been observed in isolated segments of human (Chester et al. 1989), canine

Table 5.2. Evidence of the constrictor effect of endothelin-1 on vascular smooth muscle

Preparation	Response	Reference
Anaesthetized dogs	Coronary constriction	Kurihara et al. 1989a
Anaesthetized dogs	Coronary constriction (subepicardial > subendocardial)	Clozel and Clozel 1989
Isolated rat hearts	Coronary constriction	Fukuda et al. 1989
Isolated rabbit heart	Coronary constriction	Karwatowska-Prokopczuk and Wennmalm 1990
Isolated human coronary arteries	Constriction	Chester et al. 1989
Isolated porcine coronary arteries	Constriction	Yanagisawa et al. 1988a
Isolated dog coronary arteries	Constriction	Cocks et al. 1989
Isolated rat aorta	Constriction	Sakata et al. 1989
Isolated rat aorta	Constriction	Auget et al. 1988
Isolated jugular and mesenteric veins (rabbit)	Constriction	D'Orleans-Juste et al. 1989
Aortic strips (rabbit)	Constriction	Marsden et al. 1989
Renal arteries (rat, in vivo)	Constriction	Lopez-Farre et al. 1989
Arterioles (hamster cheek pouch)	Constriction	Brain et al. 1988
Mesenteric artery (rabbit)	Constriction	Eglen et al. 1989
Basilar artery (dog, in vivo)	Constriction	Asano et al. 1989
Isolated mesenteric arteries (rat)	Constriction	Tabuchi et al. 1989
Isolated carotid arteries (dog)	Constriction	Rubanyi 1988

This table does not contain an exhaustive list of preparations used to demonstrate the contractile effect of endothelin-1 on the vasculature. Instead it is meant to show the lack of species specificity

(Cocks et al. 1989) and porcine (Yanagisawa et al. 1988a) coronary arteries, in anaesthetized dogs (Kurihara et al. 1989a, b; Clozel and Clozel 1989) and in isolated perfused rat hearts (Fukuda et al. 1989).

Non Vascular Smooth Muscle

The stimulant effect of endothelin-1 on smooth muscle contraction is not limited to vascular smooth muscle. To the contrary (Table 5.5), bronchial smooth muscle, as well as the ileum, the uterus, the taenia coli and the vas deferens all respond in this way. Some smooth muscle preparations are more sensitive than others. For example guinea pig ileum is more sensitive than either guinea pig trachea (Eglen et al. 1989) or rat vas deferens (Hiley et al. 1989). Similarly, uterine smooth muscle is more sensitive than ileal longitudinal muscle (Sakata et al. 1989; Eglen et al. 1989).

Table 5.3. Threshold concentration of endothelin-1 required to elicit a contractile response from smooth (vascular and non vascular) and cardiac muscle

Preparation	Threshold concentration of ET-1	Reference
A. Smooth Muscle		
(a) Vascular		
Rabbit microvasculature	1 pM	Brain et al. 1988
Rabbit mesenteric artery	2.5–5 pM	D'Orleans-Juste et al. 1989
Rabbit jugular and mesenteric arteries	0.5 pM	D'Orleans-Juste et al. 1989
Dog carotid artery	50 pM	D'Orleans-Juste et al. 1989
(b) Non vascular		
Guinea pig trachea	1 nM	Maggi et al. 1989a, c
Guinea pig ileum	1 nM	Hiley et al. 1989
Guinea pig trachea	1 nM	Hay et al. 1989
B. Cardiac		
Rat ventricle	20 nM	Sakata et al. 1989
Human trabeculae	50 nM	Moravec et al. 1989

Table 5.4. The concentration of endothelin-1 and of noradrenaline required to elicit fifty percent of the maximum contractile response (EC_{50}) in arterioles, venules and lymphatics

Agent	Vessel		
	Arterioles	Venules	Lymphatics
Endothelin-1	18.5 nM	18.6 nM	18.5 nM
Noradrenaline	12.3 nM	12.1 nM	No response

Data from Fortes et al. 1989. Note that although endothelin-1 has a constrictor effect on the lymphatics, noradrenaline exerts no such effect

Generally however, non vascular smooth muscle is less sensitive than vascular smooth muscle, sometimes by a factor of fifty to one hundred percent (Turner et al. 1989a, b). This difference in sensitivity is evident in Table 5.3, which shows that picomolar (10^{-12}M) concentrations of endothelin-1 are sufficient to elicit a contractile response from a variety of vascular smooth muscle preparations, but nanomolar (10^{-9}M) concentrations are needed for tracheal and ileal smooth muscle.

Table 5.5. Non vascular smooth muscles which respond to endothelin-1

Tissue	Species	Reference
Trachea	Guinea pig	Uchida et al. 1988
	Guinea pig	Maggi et al. 1989a, 1990
	Rat	Turner et al. 1989a
	Canine	Sakata et al. 1989
Bronchial	Human	Uchida et al. 1988
	Guinea pig	Maggi et al. 1989a, 1990
Lymphatic	Rat	Fortes et al. 1989
Others		
Uterine	Guinea pig	Sakata et al. 1989
Vas deferens	Guinea pig	Sakata et al. 1989
Taenia caeci	Guinea pig	Sakata et al. 1989
Ileal longitudinal	Guinea pig	Sakata et al. 1989

Characterization of Endothelin-1 Induced Contraction of Vascular and Non Vascular Smooth Muscle

Having established that vascular and non vascular smooth muscle is sensitive to endothelin-1, and *that veins are more sensitive than arteries which in turn are more sensitive than non vascular smooth muscle*, it may be appropriate to consider the general properties of the endothelin-1 induced contractile response. They can be summarized as follows. The response:

(i) *develops slowly*, and *is sustained*;

(ii) is *not* dependent on the presence of an intact endothelium – to the contrary it is enhanced, possibly because of the removal of endothelial-derived peptidases (Eglen et al. 1989) and the endothelial-derived endogenous relaxing agents, including prostacyclin and EDRF. Moreover, the response

(iii) is dose-dependent (Figure 5.2), and

(iv) is *insensitive* to tetrodotoxin (an inhibitor of Na^+ entry through the Na^+ channels which traverse excitable membranes) or Na^+ removal (Borges et al. 1989). In addition,

(v) the response is absent in preparations with permeabilized membranes (Sakata et al. 1989) and therefore *cannot* involve a direct effect of the polypeptide on the contractile proteins, and

(vi) blockade of the T- or N-type of Ca^{2+} channels with nickel ions and w conotoxin respectively (Miller 1987) does *not* affect the response, indicating that although endothelin-1 raises cytosolic Ca^{2+} (Table 5.1) any extracellular Ca^{2+} which is involved does not cross the cell membrane by way of the N- or T-type Ca^{2+} channels;*

---

* for description of N- and T-type Ca^{2+} channels see reference Nayler 1988, or Miller 1987.

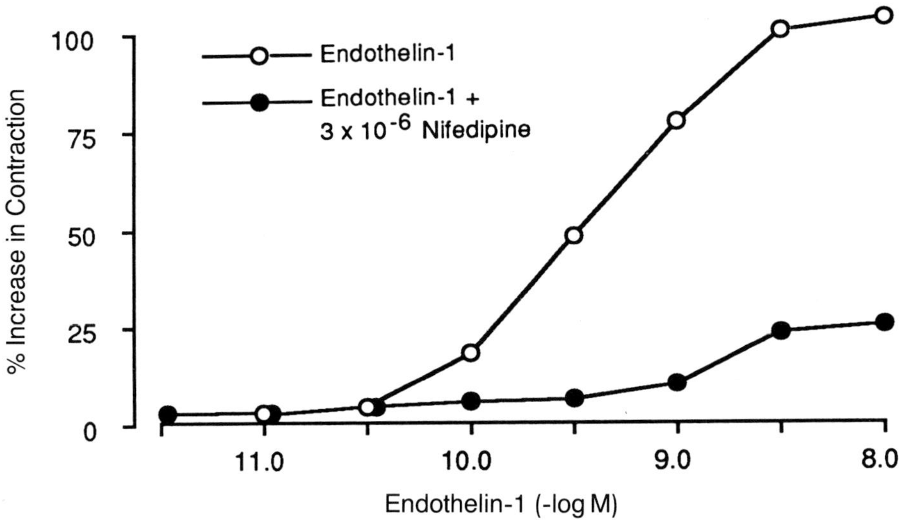

Fig. 5.2. Dose-dependent constrictor effect of endothelin-1 on isolated porcine coronary artery rings. Note that 3×10^{-6} M nifedipine attenuates the response

Table 5.6. Comparison between the contractile effects of Bay K8644 and endothelin-1

Characteristic	Bay K8644	Endothelin-1
Rate of onset	Rapid	Slow (15–20 min to reach asymptote)
Duration of response on removal of the agonist	Short (1–2 min)	Sustained (20–30 min)
Dependence on extracellular Ca^{2+}	Absolute	Not essential in many preparations
Dependence of membrane polarization	Absolute	Independent
Effect of Ca^{2+} antagonist	Complete abolition	Attenuation
Effect on cytosolic Ca^{2+}	Increased	Increased

Bay K8644 is a synthetic dihydropyridine derivative which structurally resembles nifedipine but acts in the opposite manner, by activating rather than "blocking" slow Ca^{2+} channels. It exerts its effect by binding to the dihydropyridine (nifedipine) binding sites associated with these channels (Schramm et al. 1983)

(vii) the contractile response, although accompanied by an increase in cytosolic Ca^{2+}, occurs without any change in tissue cyclic 3′5′ adenosine monophosphate (3′5′ AMP) or cyclic 3′5′ guanosine monophosphate (3′5′ GMP) (Ohlstein et al. 1989).

(viii) the response differs from that induced by the calcium agonist Bay K8644, not only in its slow rate of onset, but also in that it is only attenuated – not abolished – by calcium antagonists, and does *not* require membrane depolarization (Auget et al. 1988, and Table 5.6).

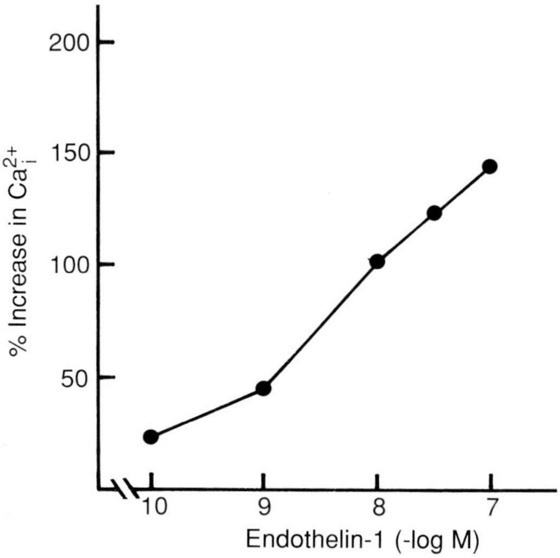

Fig. 5.3. Effect of endothelin-1 on cytosolic Ca^{2+} in isolated aortic smooth muscle cells. Note that the effect is dose-dependent. Cytosolic Ca^{2+} was measured with Fura-2. (Data from Marsden et al. 1989a)

(ix) The constrictor response is overcome by:
 (a) isoprenaline (Yanagisawa et al. 1988a);
 (b) nitroprusside (Yanagisawa et al. 1989a);
 (c) adenosine (Chester et al. 1989), and
 (d) nitroglycerine (Miller et al. 1989b), and

(x) does not involve:
 (a) α or β adrenergic;
 (b) histaminergic;
 (c) muscarinic;
 (d) serotonergic, or
 (e) opoid receptors, nor does it involve
 (f) the angiotensin converting enzyme,
 (g) angiotensin, or
 (h) vasopressin, and

(xi) is accompanied by:
 (a) a *raised cytosolic* Ca^{2+} (Figure 5.3) irrespective of whether an L-type Ca^{2+} channel blocker (e.g. nicardipine) (Mitsuhashi et al. 1989) is present;
 (b) an *increased uptake of extracellular* Ca^{2+}, (Wallnofer et al. 1989);
 (c) an *increased rate of phosphoinositol metabolism* (IP_3) (Resnik et al. 1989; Ohlstein et al. 1989; Marsden et al. 1989a) which is calcium antagonist-insensitive (Figure 5.4);
 (d) an accumulation of diacylglycerol (Griendling et al. 1989), and
 (e) protein kinase C activation (Lee et al. 1989).

Moreover, the contractile response is *blunted*, or abolished by:

(xii) the protein kinase C inhibitors, H-7 and staurosporine (Ohlstein et al. 1989).

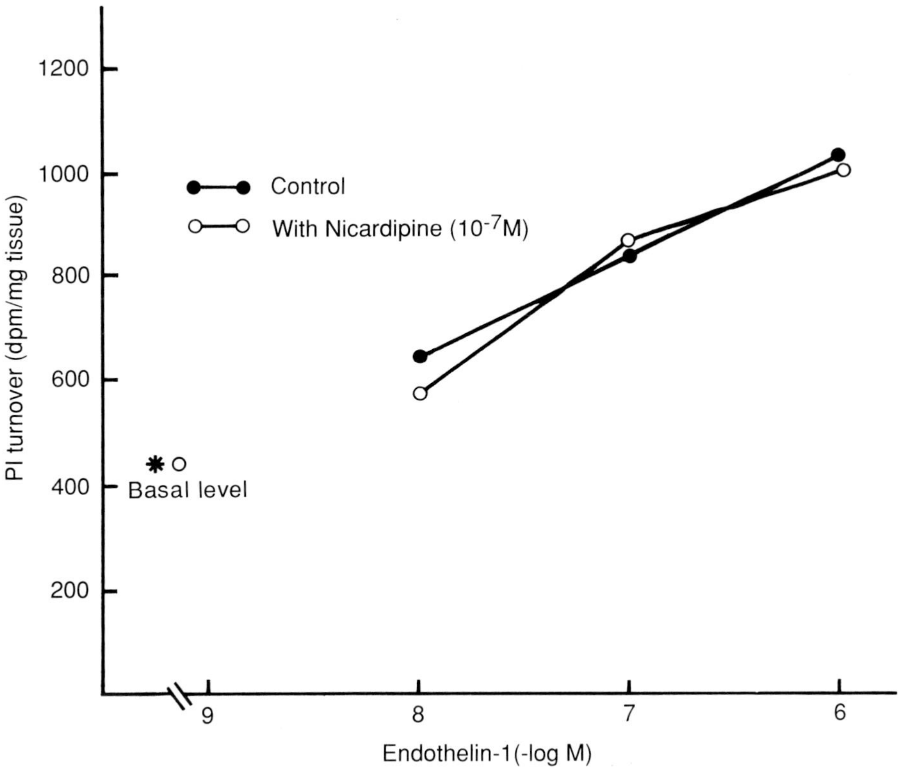

Fig. 5.4. Effect of endothelin-1 on phosphoinositol metabolism in isolated smooth muscle cells. Note that the stimulatory effect of the polypeptide is dose-dependent, and insensitive to the calcium antagonist nicardipine. (Data from Ohlstein et al. 1989)

In general, therefore, as far as the contractile effect of endothelin-1 on smooth muscle is concerned:

(i) the response has a slow rate of onset and is sustained;
(ii) some preparations are more sensitive than others (veins > arteries > non vascular);
(iii) the response is accompanied by a raised cytosolic Ca^{2+};
(iv) is attenuated by calcium antagonists and blunted by protein kinase C inhibitors; and
(v) does not involve a direct effect of endothelin-1 on the contractile proteins.

By now readers of this chapter may be wondering precisely how endothelin-1 does constrict smooth muscle. Since the response is accompanied by a raised cytosolic Ca^{2+}, an increase in Ca^{2+} influx, and is sometimes attenuated by calcium antagonists (Wallnofer et al. 1989; Yanagisawa et al. 1988a, and Figure 5.2) the simplest explanation would be that the peptide has a direct agonist-like effect on the

L-type calcium channels. There are a variety of reasons why such an explanation is too simplistic. Before these reasons are discussed, however, it may be useful to describe the effect of endothelin-1 on cardiac muscle because here, too, an increase in contractile force is accompanied by a raised cytosolic Ca^{2+} (Hirata et al. 1989a, e) and stimulation of phosphoinositol metabolism (Vigne et al. 1989).

The Positive Inotropic Effect of Endothelin-1 on Cardiac Muscle

Evidence of endothelin-1's ability to exert a positive inotropic effect on cardiac (atrial and ventricular) muscle is now relatively easy to find, provided that the response is not suppressed because of an accompanying coronary vasoconstriction. This can happen quite easily, because vascular smooth muscle is more sensitive to endothelin-1 than is cardiac muscle. For example smooth muscle, including that of the coronary vasculature, responds to picomolar amounts of the polypeptide, whereas nanomolar amounts are needed to evoke a positive inotropic response from cardiac muscle.

In many ways the positive inotropic effect of endothelin-1 resembles the constrictor response described in the previous section for the effect of endothelin-1 on vascular and non vascular smooth muscle because the response is slow in onset, taking as long as twenty minutes to reach asymptote. As with the smooth muscle preparations the response is sustained, and difficult to wash out. One interesting aspect of the response is that it is *not* accompanied by any change in the time to peak tension, or the time required for relaxation (Moravec et al. 1989). The inotropic response, therefore, is associated with an increase in the *maximum velocity of contraction and relaxation*. In this respect it resembles the response caused by a raised extracellular concentration of calcium ions but differs from the inotropic effect of isoprenaline, which abbreviates both contraction and relaxation. It is also quite different from the positive inotropic effect of angiotensin II, which is characterized by an unchanged velocity of shortening but a prolonged contraction. *These observations relating to the failure of endothelin-1 to alter either the time required to develop peak tension or the time required for relaxation are important because they indicate that*:

(i) *cyclic 3′5′ AMP is not involved* – otherwise time to peak tension and time required for relaxation would be abbreviated, as is the case with isoprenaline and other cyclic 3′5′ AMP-mediated inotropic responses; and

(ii) *angiotensin II cannot be involved*, otherwise the duration of each contraction would be prolonged – as happens with angiotensin II (Meulemans and Brutsaert 1988).

As far as the sensitivity of cardiac muscle is concerned it is certainly less sensitive than vascular smooth muscle, responding to nanomolar (10^{-9}) rather than picomolar (10^{-12}) concentrations (Table 5.3) of endothelin-1. Within the heart, however, there is a gradation of sensitivities, atria being more sensitive than ventricles. For example, the positive inotropic effect endothelin-1 exerts on human atria is at least seven times greater than its effect on ventricular muscle (a maximum increase in tension of 74.6 ± 14.1 percent and 9.4 ± 4.0 percent respectively (Moravec

Table 5.7. Evidence of the positive inotropic effect of endothelin-1 on cardiac muscle preparations

Preparation	Response	Reference
Guinea pig atria	Positive	Ishikawa et al. 1988a
Guinea pig atria (paced)	Positive	Eglen et al. 1989
Rat atria	Positive	Hu et al. 1988
Rat atria	Positive	Vigne et al. 1989
Ferret papillary muscle	Positive	Shah et al. 1989
Human atria	Positive	Moravec et al. 1989
Human trabeculae	Positive	Moravec et al. 1989

et al. 1989). The same difference applies to rat atria and ventricular muscle, although here the difference is less pronounced (a maximum increase of 39.9 ± 10.7 percent for atria, compared with 17.1 ± 5.9 percent for ventricular muscle).

In general, therefore, as far as the positive inotropic effect of endothelin-1 is concerned:

(i) the response is not species specific (Table 5.7);
(ii) it is accompanied by a decrease in resting tension (Moravec et al. 1989);
(iii) it is dose dependent;
(iv) atria are more sensitive than ventricles;
(v) the response is not dependent on an altered frequency of contraction (Moravec et al. 1989; Eglen et al. 1989); and
(vi) it is not blocked by adrenergic, histaminergic, or serotonergic antagonists (Ishikawa et al. 1988a). Also
(vii) angiotensin II appears *not* to be involved;
(viii) the response is accompanied by an increase in the inward Ca^{2+} current (I_{si}) (Ishikawa et al. 1988a),
(ix) is attenuated by calcium antagonists (Ishikawa et al. 1988a), and characteristically,
(x) is slow in onset and sustained; and
(xi) is accompanied by a raised cytosolic Ca^{2+} (Hirata et al. 1989e) and enhanced phosphoinositide metabolism (Vigne et al. 1989).

Does Endothelin-1 Exert a Positive Chronotropic Effect on the Heart?

At the moment this is a surprisingly difficult question to answer, because opinion is divided. Ishikawa et al. (1988b) detected a positive chronotropic response in guinea pig spontaneously beating right atrial preparations but at least two other groups (Hu et al. 1988a, and Moravec et al. 1989) failed to find any evidence of a positive chronotropic response from rat right atrial preparations. Maybe there is a species difference or perhaps in the rat atrial studies the development of tachyphylaxis obscured the response. On a purely theoretical basis a positive chronotropic response might have been anticipated, because many other agents which increase the slow inward Ca^{2+} current (I_{si}) do have a positive chronotropic effect on the heart.

Summary

Endothelin-1 is a potent constrictor of vascular and non vascular smooth muscle. It also has a positive inotropic effect on the heart, constricts lymphatic vessels and, at least in the guinea pig, has a positive chronotropic effect.

As far as its potency is concerned the sensitivity of vascular smooth muscle to endothelin-1 > non vascular smooth muscle and in the heart atria are more sensitive than the ventricles.

The contractile effect of endothelin-1, and its effect on the myocardium, is dose-dependent, develops slowly, is sustained, and does not involve angiotensin II, vasopressin, α or β adrenergic, serotonergic, histaminergic or opoid receptors, or an altered level of 3′5′ AMP or 3′5′ GMP. In both tissues, however, the contractile effect is accompanied by a raised cytosolic Ca^{2+}.

Chapter 6

The Biochemical Events Involved in the Contractile and Positive Inotropic Activity of Endothelin-1

> *"I dogmatize and am contradicted, and in this conflict of opinions and sentiments I find delight"*
> SAMUEL JOHNSON

The previous chapter concluded with the statement that *"the contractile effect of endothelin-1 is accompanied by a raised cytosolic Ca^{2+}*. Whilst it is a truism to say that it is the raised cytosolic Ca^{2+} (Figure 6.1) which is directly responsible for the contractile and positive inotropy of this polypeptide the cascade of events which culminates in this increase are complex. The initial, or primary, event, however, is the binding of the polypeptide to specific, membrane-located high affinity binding

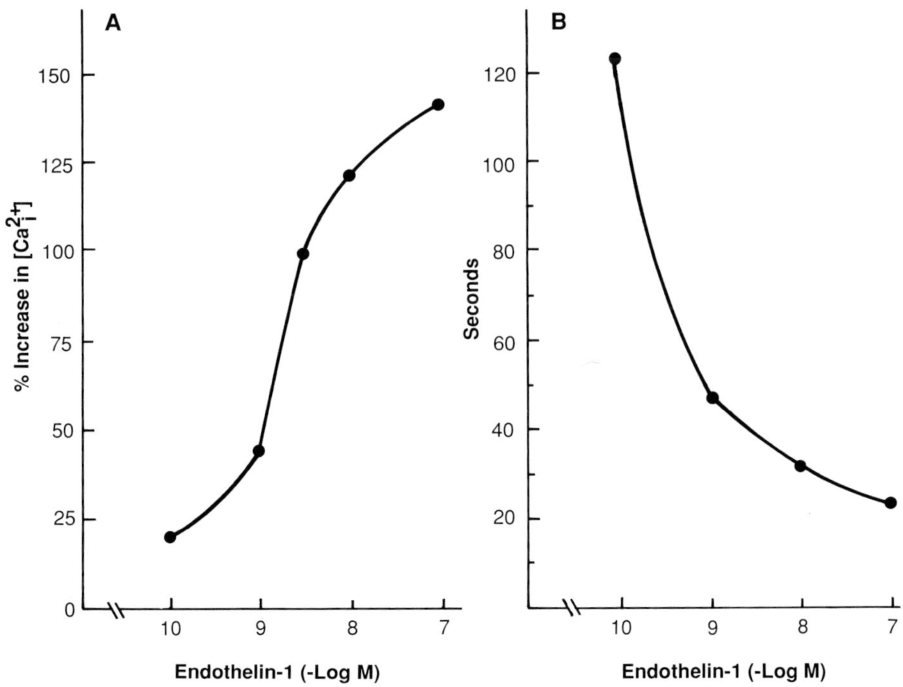

Fig. 6.1. Dose-dependency of the time to peak tension (B) and the concentration-response curves (A) for the endothelin-1 induced increase in cytosolic Ca^{2+} (Ca_i^{2+}) measured with Fura-2 in vascular smooth muscle cells. (Data from Marsden et al. 1989a)

sites – in much the same way as occurs for other vasoactive polypeptides, including angiotensin II and vasopressin. The properties, size and location of these endothelin receptors are described in the next chapter (Chapter 7). The present chapter is concerned only with the sequence of events which are triggered by the binding of the endothelins to their specific receptor(s), and which culminates in the raised cytosolic Ca^{2+} which ultimately dictates the magnitude of the contractile and inotropic responses.

Endothelin – and Vascular Smooth Muscle

Origin of the raised cytosolic Ca^{2+}

When Yanagisawa and his co-workers (Yanagisawa et al. 1988a) first studied the pharmacological properties of endothelin-1 they found that its ability to constrict porcine coronary arteries is:
(i) completely reversed by isoprenaline and glyceryl trinitrate;
(ii) resistant to α adrenergic, H_1-histaminergic, serotonergic, lipoxygenase and cyclooxygenase blockers;
(iii) markedly attenuated by the calcium antagonist, nicardipine, and
(iv) dependent upon the presence of extracellular Ca^{2+}.

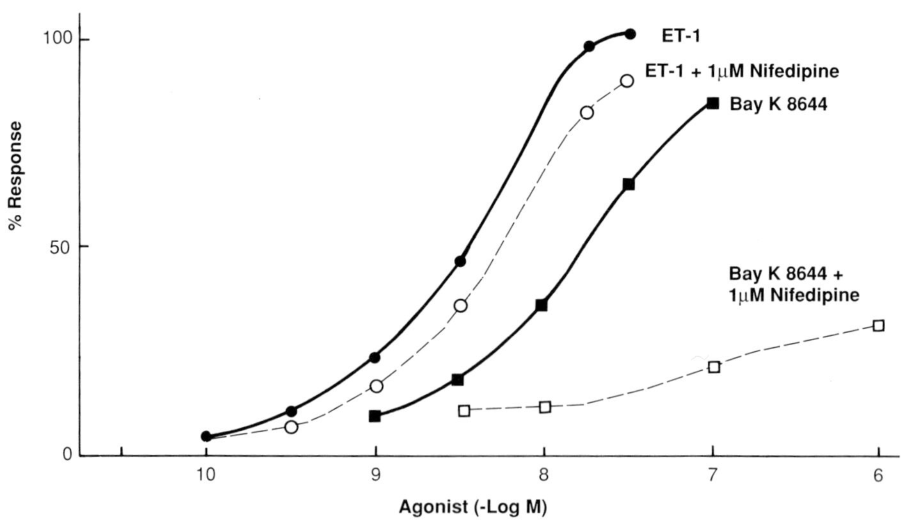

Fig. 6.2. Effect of 1 μM nifedipine on the contractile effect of endothelin-1 and the Ca^{2+} agonist Bay K8644 on isolated denuded rat aortic preparations superfused with Krebs-Henseleit buffer containing 2.5 mM Ca^{2+}. Temperature was maintained at 37°C. For the Bay K8644 series the preparation was partially depolarized by increasing the KCl content of the buffer from 4.7 to 12 nM five minutes before beginning the cumulative dose-response curves. Note that whereas nifedipine had a marked effect on the constrictor effect of Bay K8644 the effect of endothelin-1 was only slightly suppressed

Table 6.1. Sensitivity of various vascular and non vascular muscle preparations to endothelin-1

Tissue	Response	Reference
Cardiac Tissue		
Rat atria	Blocked by D888	Vigne et al. 1989
Guinea pig atria	Attenuated by nicardipine	Ishikawa et al. 1988a
Vascular Smooth Muscle		
Jugular veins	Insensitive to nicardipine	D'Orleans Juste et al. 1989
Rat aorta	Blocked by verapamil and nifedipine	Eglen et al. 1989
Rat aorta	Blocked by nitrendipine and nifedipine	Borges et al. 1989
Rat aorta	Dose-response curve shifted to the right by diltiazem, D600 and nifedipine	Auget et al. 1988
Rabbit aorta	Relatively insensitive to diltiazem, nicardipine, nifedipine	Ohlstein et al. 1989
Porcine coronary artery	Sensitive to nicardipine	Yanagisawa et al. 1988a
Dog coronary artery	Sensitive to nitrendipine	Kurihara et al. 1989a
Human coronary artery	Sensitive to nifedipine	Chester et al. 1989
Renal artery	?	Firth et al. 1988
Non vascular smooth muscle		
Rat trachea	Insensitive to nicardipine	Turner et al. 1989b

? denotes unknown

These observations might have been regarded as providing the evidence necessary to support the notion that endothelin-1 might be the hitherto elusive but naturally occurring endogenous agonist of the voltage-dependent Ca^{2+} channels which was, and still is, being so keenly sought after. The search for such a substance is a logical aftermath of the synthesis of the synthetic calcium channel agonist, Bay K8644 (Schramm et al. 1983). However, although the contractile effect of endothelin-1 is accompanied by an enhanced influx of Ca^{2+} (Wallnofer et al. 1989), and although it usually (Borges et al. 1989) but not always (Marsden et al. 1989a) requires extracellular Ca^{2+}, the *direct* activation of the slow calcium channels has now been shown *not* to be involved in the vasoconstrictor response. There are several reasons for reaching this conclusion:

(i) even when relatively large concentrations of calcium antagonists are used, the dose-response curves for the vasoconstrictor effect of endothelin-1 is only shifted to the right, and not abolished as happens for the synthetic agonist Bay K8644 (Auget et al. 1988, and Figure 6.2, and Table 6.1);

(ii) the constrictor effect of endothelin-1 on some smooth muscle preparations is relatively insensitive to Ca^{2+} withdrawal. Rat and rabbit aorta, and even human coronary arteries can behave in this way (Table 6.2), and

Table 6.2. Sensitivity to extracellular Ca^{2+} of endothelin-1 induced response in smooth and cardiac muscle

Preparation	Sensitivity to extracellular Ca^{2+}	Reference
Porcine coronary artery	Yes	Yanagisawa et al. 1988a
Rat trachea	Yes	Turner et al. 1989b
Rat atria	Yes	Vigne et al. 1989
Rat portal vein	Yes	Borges et al. 1989

(iii) endothelin-1 is incapable of displacing calcium antagonists (diltiazem, isradipine and verapamil, for example) from their specific binding sites. These binding sites are an integral part of the L-type calcium channels (Nayler 1988) and they provide the major site of action of the calcium agonists and antagonists. If the endothelins had been acting directly on these calcium channels then they should have been able to displace the previously bound channel agonists or antagonists – unless they were interacting with an entirely different binding site.

The problem, then, is to explain how endothelin-1 increases cytosolic Ca^{2+} in vascular smooth muscle cells. Various possibilities exist, including:

(i) an inhibition of Ca^{2+} efflux. *This does not occur*, however. To the contrary, Ca^{2+} efflux is enhanced (Wallnofer et al. 1989; Marsden et al. 1989a). Alternatively,

(ii) an enhanced entry of Ca^{2+} in exchange for Na^+, by way of a *direct* effect on the Na^+:Ca^{2+} exchanger could be involved. This possibility also can probably be excluded, since the peptide has no effect on the kinetics of the exchanger nor is its effect dependent on the presence of extracellular Na^+ (Criscione et al. 1989). However, enhanced Ca^{2+} entry by way of the Na^+:Ca^{2+} exchanger could be involved, since in some tissues (Gardner et al. 1989b; Simonson and Dunn 1990) the endothelins stimulate the Na^+:H^+ exchanger. Such an effect increases the amount of Na^+ which is available for exchange with Ca^{2+} without altering the kinetics of the exchanger.

(iii) an effect on the Na^+K^+ ATPase could be involved, with an inhibition of the enzyme resulting in an increased intracellular availability of Na^+ for exchange with Ca^{2+} by way of the Na^+:Ca^{2+} exchanger. An inhibitory effect on the Na^+K^+ ATPase cannot be an essential requirement for the contractile effect of endothelin-1, however, because this inhibitory effect only occurs at concentrations which are about one thousand times greater than that required for vasoconstriction (Meyer-Lehnert et al. 1989a, b).

(iv) Ca^{2+} may be being mobilized from the intracellular stores. This seems to be an acceptable hypothesis, because recent experiments in which the Ca^{2+}-sensitive dye Fura-2 has been used to monitor the changes in cytosolic Ca^{2+} caused by endothelin-1 (Mitsuhashi et al. 1989; Simpson and Ashley 1989; Meyer-

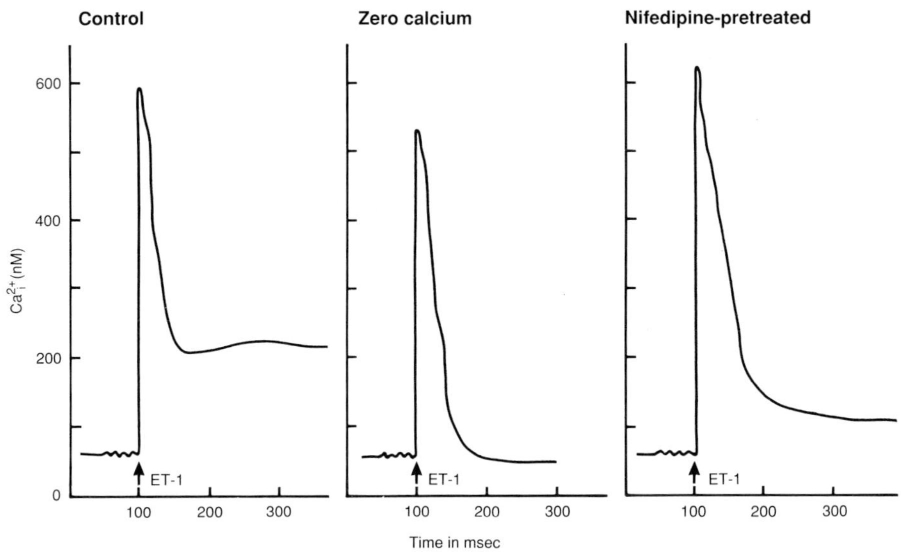

Fig. 6.3. Effect of 1 nM endothelin-1 on cytosolic Ca^{2+} in isolated aortic smooth muscle cells, as measured with Fura-2. In the first panel the cells were incubated in medium containing 1.3 nM Ca^{2+}. In the middle panel Ca^{2+} was withdrawn before the endothelin-1 was added. In the third panel Ca^{2+} was returned but the cells were incubated together with 10 μM nifedipine before 1 nM endothelin-1 was added. Different cells were used for each series, to avoid "down regulation" of the receptors

Lehnert et al. 1989a, b; Wallnofer et al. 1989) have shown not only that endothelin-1 increases cytosolic Ca^{2+} but also that the increase is:

(a) dose-dependent (Figure 6.1);
(b) not dependent on extracellular Ca^{2+} (Figure 6.3), and
(c) biphasic (Figure 6.3 first panel). There is an initial sharp spike (which is caffeine- and ryanodine-sensitive, indicating that the Ca^{2+} which is involved originated from the sarcoplasmic reticulum) followed (first panel of Figure 6.3) by a calcium antagonist sensitive plateau (Simpson and Ashley, 1989). The sensitivity of the plateau phase to calcium antagonists implies that the Ca^{2+} which is involved in this facet of the response entered by way of the voltage-sensitive calcium-selective channels, since it is these channels which are sensitive to the calcium antagonists.

Hence, in vascular smooth muscle the raised cytosolic Ca^{2+} which facilitates the contractile effect of endothelin-1 originates from at least two pools:

(i) release from the sarcoplasmic reticulum;
(ii) extracellular Ca^{2+} entering either through the indirectly activated calcium channels or by way of the Na^+:Ca^{2+} exchanger.

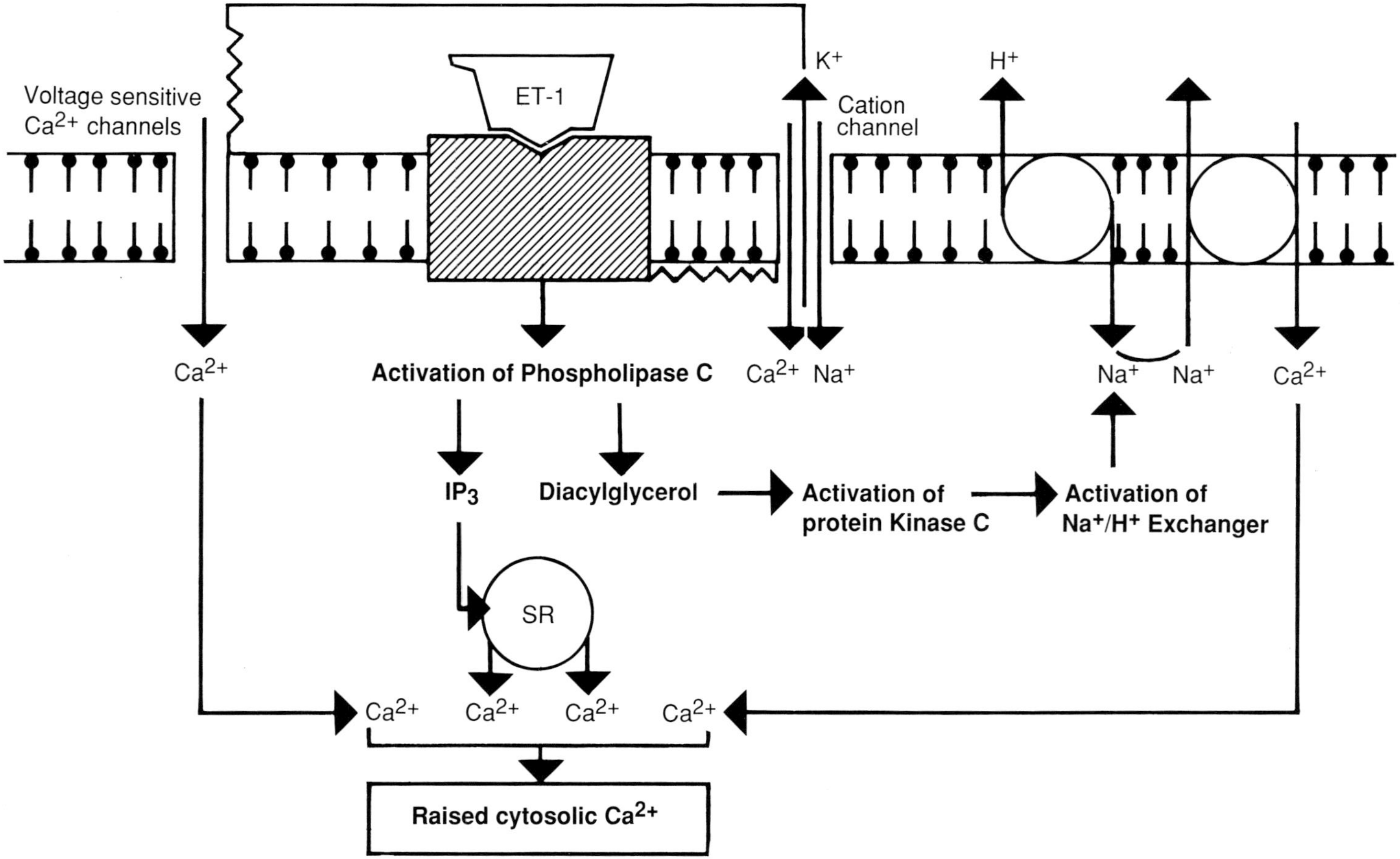

Fig. 6.4. Schematic representation of the possible events involved in the endothelin-1 induced increase in cytosolic Ca^{2+}

Mechanisms Involved in the Mobilization of Intracellular Calcium and Enhanced Slow Channel Ca^{2+} Influx

Mobilization of intracellular Ca^{2+}

The key to our understanding of the events which are involved in endothelin-1's ability to mobilize intracellular Ca^{2+} was provided by the observation that endothelin-1 stimulates the phospholipase C enzyme (Resnik et al. 1989; Simonson et al. 1989). Activation of phospholipase C is a primary step because it allows for the hydrolysis of membrane-located inositol lipids (PI) (namely phosphatidylinositol and its phosphorylated derivatives phosphatidylinositol 4 – phosphate (PIP) and 1-phosphatidyl-D-myoinositol, 4,5-biphosphate (PIP2). This hydrolysis releases two compounds, inositol triphosphate (IP_3) and diacylglycerol (DG) (Figure 6.4). IP_3 migrates into the cytoplasm and is responsible for the rapid mobilization of intracellular Ca^{2+}.

Measurement of inositol phosphates in vascular smooth muscle cells has shown that endothelin-1 does cause a *rapid* (Figure 6.5) formation of inositol triphosphate (Marsden et al. 1989a), that this response does not require extracellular Ca^{2+}, is

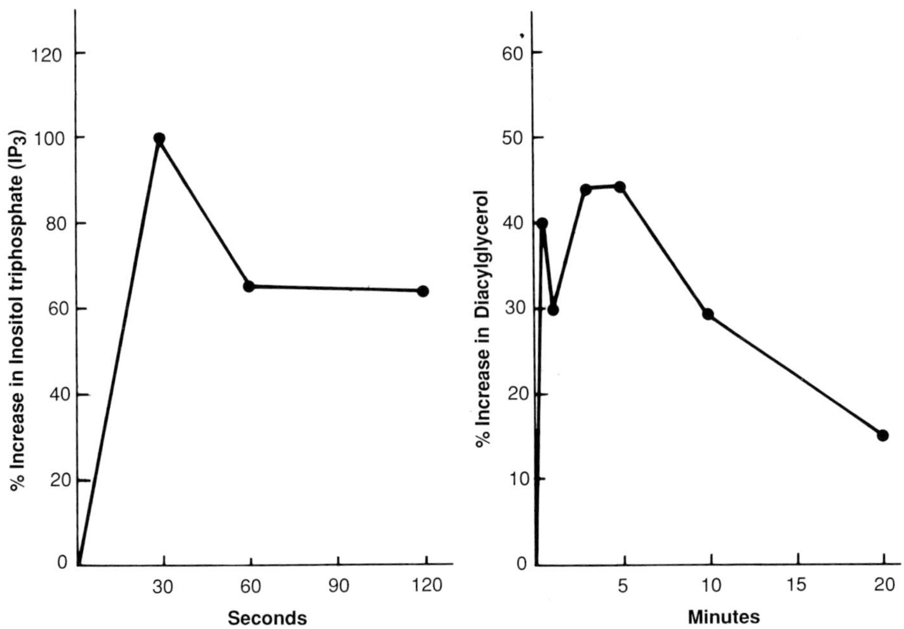

Fig. 6.5. Time course for the effect of endothelin-1 (100 nM) on the formation of inositol triphosphate (IP_3) and diacylglycerol (DA) in vascular smooth muscle cells. Data are expressed as percentage increase over the basal levels. Note the rapid effect on IP_3, and the slower effect on DA generation. In the case of DA the response is characteristically biphasic with an early peak at around 5 seconds followed by a second peak two – three minutes later. Note also the rapidity with which the IP_3 levels increase. (Data from Kasuya et al. 1989a, b, and Griendling et al. 1989)

dose-dependent (Kasuya et al. 1989a, b) and is not accompanied by altered tissue levels of cyclic 3'5' AMP (Sugiura et al. 1989a). These criteria all fit the characteristics of the endothelin-induced increase in cytosolic Ca^{2+}. Since it is the spike component of the Ca^{2+}-signal (Figure 6.3) which is caffeine sensitive (Kai et al. 1989), it is likely that this fast spike reflects the IP_3-induced release of Ca^{2+} from the sarcoplasmic reticulum.

The hydrolysis of inositol lipids also releases diacylglycerol, as well as IP_3 (Figure 6.4). The diacylglycerol is retained in the membrane where its primary signalling role is the activation of a phospholipid-dependent protein kinase C, an enzyme which, amongst other things, activates the Na^+:H^+ exchanger. Activation of the Na^+:H^+ exchanger provides one of the possible routes for Ca^{2+} entry into vascular smooth muscle cells, the Na^+ exchanging for Ca^{2+} by way of the Na^+:Ca^{2+} exchanger (Figure 6.4).

Since endothelin-1:

(i) stimulates the accumulation of diacylglycerol in vascular smooth muscle (Resnik et al. 1989; Muldoon et al. 1989 and Griendling et al. 1989);
(ii) enhances the activity of the Na^+:H^+ exchanger (Simonson et al. 1989), and
(iii) its contractile effect is inhibited by inhibitors of the protein kinase C enzyme (Sugiura et al. 1989a)

it seems probable that *at least two of the products formed by the hydrolysis of membrane-located inositol lipids – inositol triphosphate and diacylglycerol – both contribute to the mechanisms whereby endothelin-1 raises cytosolic Ca^{2+} by a calcium-antagonist insensitive pathway.*

Enhanced Slow Channel Ca^{2+} Entry

The endothelin-1-induced increase in cytosolic Ca^{2+} (Simpson and Ashley 1989) in vascular smooth muscle (Figure 6.3) clearly consists of two components – a calcium antagonist insensitive, caffeine sensitive spike, followed by a calcium antagonist sensitive plateau.

The preceding few paragraphs have described the evidence which has led to the conclusion that endothelin-1 mobilizes Ca^{2+} which is responsible for the early rise in cytosolic Ca^{2+}, and that the phospholipase C-induced hydrolysis of the membrane-located inositol lipids provides the trigger for the response. At first sight the suggestion of there being a calcium antagonist-sensitive component of the Ca^{2+} signal may seem puzzling, particularly since a *direct* effect of the polypeptide on the Ca^{2+} channels has been excluded as a major cause of the constrictor effect of endothelin-1. However,

(i) in many preparations (Table 6.1 and Figure 6.2) the dose response curve for the constrictor effect of endothelin-1 is shifted to the right, and
(ii) endothelin does enhance the slow inward Ca^{2+} current (Wallnofer et al. 1989; Silberg et al. 1989).

The answer to this puzzle probably lies in the fact that endothelin-1, by activating non-selective transmembrane cation carrying channels causes some small degree of membrane depolarization – small but of sufficient magnitude to *indirectly* activate the Ca^{2+} channels (Van Renterghem et al. 1988). In this case, part of the rise in cytosolic Ca^{2+} could be attributable to Ca^{2+} entry by way of a calcium antagonist-sensitive pathway which has been indirectly activated.

In summary, therefore, as far as vascular smooth muscle is concerned, the endothelin-induced increase in cytosolic Ca^{2+} involves two pathways:

(i) phospholipase C-induced increase in phosphoinositol metabolism, giving rise to increased levels of inositol triphosphate and diacylglycerol which in turn release Ca^{2+} from the sarcoplasmic reticulum and indirectly enhance Ca^{2+} entry through the Na^{+}:Ca^{2+} exchanger; and

(ii) Ca^{2+} entry through indirectly-activated slow Ca^{2+} channels.

Non Vascular Smooth Muscle.

The endothelins also increase tension development in non vascular smooth muscle – including the ileum, oesophageal muscularis mucosa, uterus and trachea – but the sensitivity of these preparations is less than that of their vascular counterparts. For example rat trachea is approximately one hundred times *less* sensitive to endothelin-1 than rat aorta. As far as the relative contractile effects of the endothelial isoforms are concerned relatively little information is available concerning their effect on non vascular smooth muscle but if rat uterine muscle is anything to go by, endothelin-1 is more effective than endothelin-3 (Figure 6.6).

Tracheal Smooth Muscle

To some extent the contractile effect of endothelin-1 on this type of non vascular smooth muscle resembles that described in the preceding paragraphs for vascular smooth muscle. Both tissues contain high affinity binding sites for the endothelins (Chapter 7), and in both cases the contractile response develops slowly and is maintained. The response is dose-dependent but in tracheal muscle it is relatively insensitive to the *calcium antagonists* (Tables 6.1 and 6.3), and removing extracellular Ca^{2+} only abbreviates and reduces, but does not abolish the response (Table 6.3) to any great extent. Hence, as far as tracheal smooth muscle is concerned the contractile effect of the endothelins, and more particularly that of endothelin-1, must depend largely upon the mobilization of internally stored Ca^{2+}, with a small component coming from an extracellular source. This extracellular Ca^{2+} cannot be entering by way of the slow Ca^{2+} channels, otherwise the contractile response would be sensitive to calcium antagonists, and this does not happen (Tables 6.1 and 6.3). In this respect, therefore, the effect of endothelin-1 on tracheal smooth muscle contrasts with its effect on vascular preparations where (Table 6.1) the constrictor response is usually modified by the calcium antagonists.

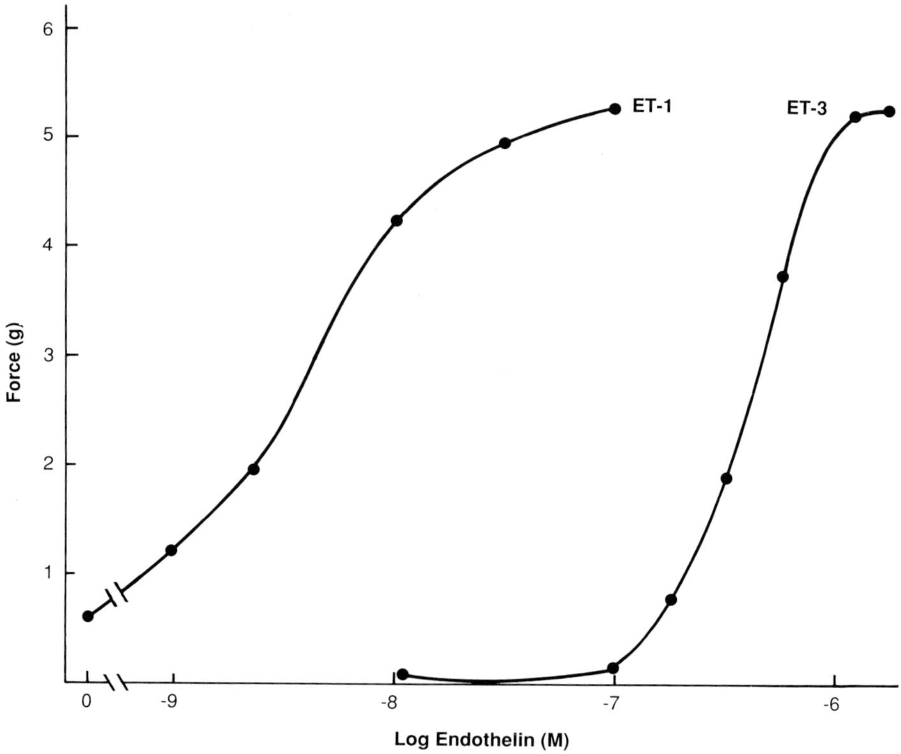

Fig. 6.6. Effect of endothelin-1 (ET-1) and endothelin-3 (ET-3) on tension developed by isolated rat uterine muscle superfused with oxygenated (95% O_2 + 5% CO_2) Krebs-Henseleit buffer at 31 °C. The doses of the endothelins were applied cumulatively.(Data from Bousso-Mittler et al. 1989)

Table 6.3. Contractile effect of 1 μM endothelin-1 on rat tracheal smooth muscle

Perfusion buffer	Increase in contractile response (g)	Duration of response (minutes)
O Ca^{2+}	0.86	17
2.5 mM Ca^{2+}	2.2	57
2.5 mM Ca^{2+} + 1 μM nicardipine	2.1	57

Rat tracheal rings were superfused with Ca^{2+}-free and Ca^{2+}-containing Krebs-Henseleit buffer, with and without 0.1 μM nicardipine. (Data from Turner et al. 1989b)

Whether endothelin-1 stimulates phosphoinositol metabolism in tracheal smooth muscle does not seem to have been investigated, as yet. Accordingly, no information is available as to whether an increased level of inositol triphosphate (IP_3) triggers the release of Ca^{2+} from the internal stores in that tissue, but it seems likely that this will be the case. Nor is it known, as yet, whether status asthmaticus increases the sensitivity of tracheal (and bronchial) smooth muscle to the endothelins.

Uterine Smooth Muscle

The effect of endothelin-1 on uterine smooth muscle almost certainly involves the phosphoinositols, since this polypeptide produces a time- and dose-dependent increase in phosphoinositol hydrolysis in this tissue (Bousso-Mittler et al. 1989). Between 20 and 40 nM endothelin-1 causes a fifty percent increase in hydrolysis, and at this concentration the polypeptide causes a *sustained, monophasic increase in tension development which is dependent on extracellular Ca^{2+}, but calcium antagonist insensitive* (Kozuka et al. 1989). *At lower concentrations (1 nM) endothelin-1* has an entirely different effect – *it produces spontaneous contractions* which sometimes, but not always, are accompanied by incomplete relaxation. This phase of the response also requires extracellular Ca^{2+}, but is *calcium antagonist sensitive*. Hence, as can be seen from Figure 6.7, the contractile effect of endothelin-1 on uterine smooth muscle is complex but can be divided into two processes:

(i) a calcium antagonist-*sensitive* increase in the frequency and amplitude of spontaneous contractions which is dependent upon extracellular Ca^{2+} and is the predominant response at low concentrations of endothelin-1, and

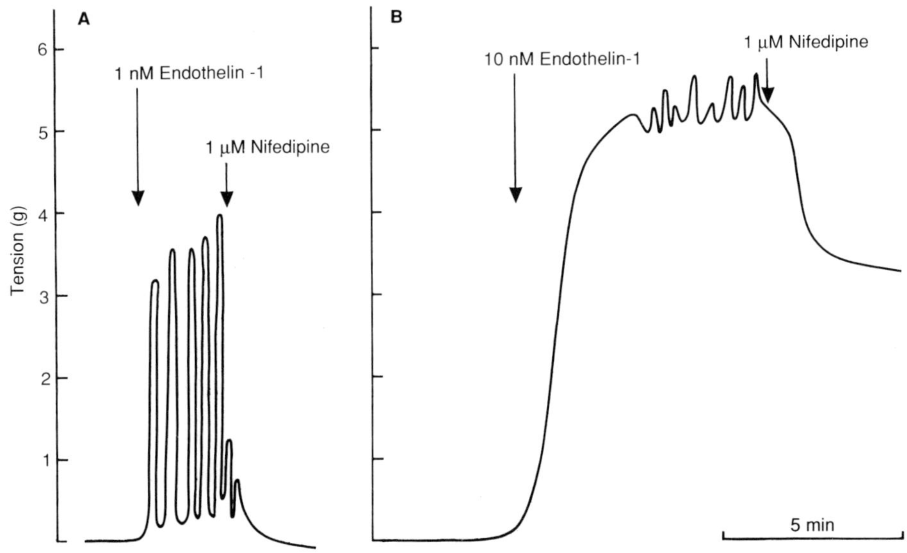

Fig. 6.7. Effect of 1 μM nifedipine on the contractile response of rat uterine muscle to (A) 1 nM and (B) 10 nM endothelin-1. Note that the lower concentration of the polypeptide caused spontaneous contractions which were nifedipine *sensitive*, whereas the larger dose (10 nM) caused a sustained rise in tension which was nifedipine *insensitive*. Both responses required extracellular Ca^{2+}

(ii) a calcium antagonist-*insensitive* sustained monophasic contraction which requires extracellular Ca^{2+} and which predominates at the higher concentrations of endothelin-1 which also stimulate phosphoinositol metabolism.

One way of explaining these events is to assume that the Ca^{2+} ions which contribute to the spontaneous contractions seen at relatively low endothelin-1 concentrations enter by way of the Ca^{2+} channels. Activation of these channels must again be indirect, because as described for vascular smooth muscle, endothelin-1 does not interact directly with the calcium antagonist binding sites which are associated with these channels (Kozuka et al. 1989). Activation of these channels could result from membrane depolarization caused by the activation of non-selective, cation carrying channels – as Van Renterghem et al (1988) described for vascular smooth muscle, or it could involve an inositol triphosphate activation of the calcium channel (Berridge and Irvine 1989).

The majority of the Ca^{2+} ions that mediate the sustained contraction of non vascular smooth muscle are almost certainly released from an intracellular store, the Ca^{2+} content of which is probably in equilibrium with that in the extracellular fluid. This release of internally stored Ca^{2+} occurs at relatively high endothelin-1 concentrations, and since these high concentrations activate phosphoinositol hydrolysis it is likely that one of the products of this hydrolysis – probably phosphoinositol triphosphate – triggers the release process.

Endothelin-1 Induced Increase in Cytosolic Ca^{2+} in Cardiac Myocytes

The positive inotropic effect of endothelin-1 on cardiac (atrial and papillary) muscle can be summarized as follows:

(i) it is not dependent upon an altered frequency of contraction;
(ii) the response is relatively slow in onset, and sustained;
(iii) it is accompanied by an increase in the inward Ca^{2+} current (Ishikawa et al. 1988a) and cytosolic Ca^{2+} (Hirata et al. 1989a);
(iv) it requires extracellular Ca^{2+}, and is attenuated by the calcium antagonists (Ishikawa et al. 1988a). In addition
(v) it is not mediated by protein kinase C but is accompanied by an enhanced rate of phosphoinositol hydrolysis (Vigne et al. 1989; Ambar et al. 1989a);
(vi) the response is either attenuated or abolished by drugs (caffeine and ryanodine) which interfere with Ca^{2+} release from the sarcoplasmic reticulum (Vigne et al. 1989), and
(vii) as with vascular and non vascular smooth muscle preparations, adding endothelin-1 to isolated cardiocytes causes a biphasic increase in cytosolic Ca^{2+}, as indicated by the Fura-2 signal. There is an initial, rapidly developing spike, which is calcium antagonist-insensitive, followed by a secondary calcium antagonist-sensitive plateau phase (Hirata et al. 1989e) – a pattern of response which is almost identical with that obtained for vascular smooth muscle.

A logical explanation for these observations is that part of the Ca^{2+} which mediates the inotropic effect of endothelin-1 on cardiac muscle is derived from an internal store, probably the sarcoplasmic reticulum. Maintenance of this store must depend upon retrieval of Ca^{2+} from the cytosol, as well as Ca^{2+} entry through the voltage-sensitive channels. Since endothelin-1 does not interact directly with these channels (Chapter 7) their activation by the polypeptide must again involve an indirect process, as already described for smooth muscle. At relatively low doses of endothelin-1, Ca^{2+} entry through the indirectly-activated Ca^{2+} channels probably provides the major amount of Ca^{2+} needed for the inotropic response but at higher doses, the concommitant activation of phosphoinositol hydrolysis generates inositol triphosphate (IP_3), a known trigger for Ca^{2+} release. It also generates diacylglycerol (DA), a known activator of protein kinase C. In smooth muscle, protein kinase C probably contributes to the overall response (Griendling et al. 1989; Lee et al. 1989; Muldoon et al. 1989; Ohlstein et al. 1989) but in cardiac muscle protein kinase C activation appears not to be of such importance (Vigne et al. 1989).

Hence, in cardiac and vascular smooth muscle cells, and in some but not all non vascular smooth muscle, there are at least two and possibly three pathways for the supply of Ca^{2+} needed to mediate the contractile and inotropic effects of endothelin-1. One route involves Ca^{2+} entry through the indirectly activated voltage sensitive Ca^{2+} channels, another involves Ca^{2+} entry in exchange for Na^+, and the other, release of internally stored Ca^{2+}. The relative contribution of each of these sources of Ca^{2+} to the contractile response is dependent, to some extent, on the concentration of endothelin-1 that is present, and the tissue type. Naturally, it must be the relative importance of each of these two sources of Ca^{2+} which determines whether the endothelin-1 response is calcium antagonist-sensitive or -insensitive. Irrespective of which tissue is involved, however, the response is initiated (Figure 6.4) by the activation of specific membrane-located endothelin-1 binding sites, – or receptors. These sites are complex and are the subject of the next chapter.

Summary

1. The events involved in the contractile and positive inotropic activity of endothelin-1 are complex (Figure 6.8).
2. Irrespective of tissue type (cardiac or smooth muscle) the ability of endothelin-1 to enhance tension development involves its ability to raise cytosolic Ca^{2+}.
3. This cytosolic Ca^{2+} is both of intra- and extracellular origin.
4. The endothelin-1 induced increase in cytosolic Ca^{2+} includes a component of Ca^{2+} which enters through indirectly activated Ca^{2+} channels. In addition Ca^{2+} is mobilized from the intracellular stores within the sarcoplasmic reticulum, or it may enter in exchange for Na^+.
5. This mobilization of the intracellular Ca^{2+} is probably triggered by the endothelin-induced activation of phosphoinositol metabolism.

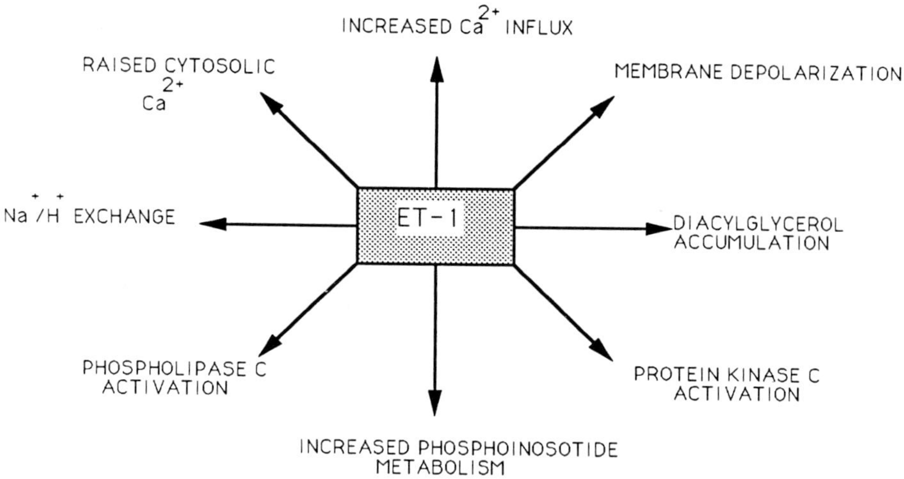

Fig. 6.8. Schematic representation of the factors involved in the effect of ET-1 with respect to Ca^{2+}

6. The relative contribution of each of these Ca^{2+}-supply pathways in both the functioning of the tissue and the mode of action of the endothelins will determine, in part at least, the sensitivity of the endothelin-induced response to calcium antagonists.

Chapter 7

Endothelin Receptors: Their Distribution and Properties

"But do you believe all that is written," asked Lady Honstanton. "I do nowadays"' Lord Illingworth replied. "It is only the unreadable that occurs."
OSCAR WILDE in: A Woman of No Importance

In their original endothelin paper Yanagisawa and his colleagues (Yanagisawa et al. 1988a) concluded that the polypeptide probably resembles the Ca^{2+} agonist Bay K8644 in its mode of action. Bay K8644 is the nifedipine derivative which binds directly to the dihydropyridine binding sites in the voltage-sensitive L-type Ca^{2+} channels (Schramm et al. 1983). Whereas nifedipine reduces, Bay K8644 *increases* the probability of these channels remaining in their "open" state and hence being available for Ca^{2+} influx. Yanagisawa and his co-workers originally thought that endothelin might have a similar mode of action because:

(i) structurally it shows some homology with other naturally occurring neurotoxins (for example α scorpion toxin) which directly interact with other ion conducting channels;
(ii) the constrictor effect of endothelin on the porcine coronary artery required extracellular Ca^{2+}, and
(iii) Ca^{2+} antagonists, including nicardipine, attenuate the constrictor response (Yanagisawa et al. 1988a).

Whilst the assumption that the endothelins interact directly with membrane-located binding sites is now known to be correct, experience has taught us that these binding sites are clearly distinct from those which interact with the calcium antagonists (Gu et al. 1989b–d; Miyazaki et al. 1990). Indeed, the endothelin binding sites (receptors) may not even be located in, or in any way associated with, the Ca^{2+} channels. The present chapter is intended to provide an overview of our current knowledge concerning these endothelin binding sites, particularly with respect to their tissue distribution, properties and size.

Tissue Distribution of the Endothelin Binding Sites

Once iodinated endothelin (Figure 7.1) became available many investigators turned their attention to mapping the distribution of the endothelin binding sites. Contrary to initial expectations, these membrane-located binding sites were found to be widely distributed throughout the body – and not restricted to the cardiovascular system, as was initially anticipated. Other surprising findings soon emerged –

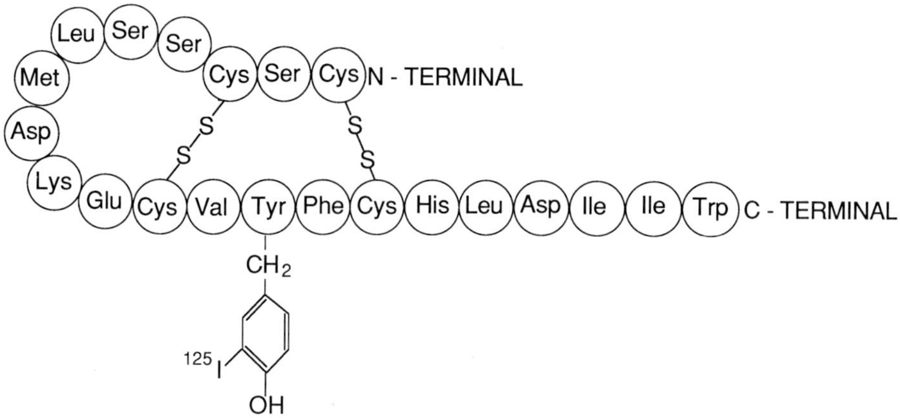

Fig. 7.1. Structure of iodinated endothelin

Table 7.1. Tissue distribution of endothelin binding sites

Organ	Distribution
A. Tissues with endothelin binding sites	
Heart	Myocytes (atrial and ventricular), atrio-ventricular nodes, coronary arteries and veins
Kidney	Glomerulus, renal papilla, renal artery and veins, interlobular and arcuate arteries, mesangial cells
Eye	Retinal capillaries, corneal endothelium, ciliary body, iris
Skin	Fibroblasts
Brain	Cerebellum (hypothalamus), pituitary gland
Lung	Bronchi, trachea, alveoli, pulmonary artery
Stomach and gut	Mucosal lining
Adrenal gland	Medulla, cortex
Vasculature	Aorta, pulmonary artery, coronary artery, abdominal aorta, renal artery, cerebral artery, renal vein, coronary vein, umbilical vein
Spleen	
Liver	
Placenta	
Vas deferens	
Uterus	
B. Tissues without endothelin binding sites	
Adipocytes	
Connective tissue	
Cartilage	
Platelets	
Red blood cells	

This list is only to illustrate the remarkable widespread distribution of the endothelin binding sites

Table 7.2. Relative distribution of endothelin binding sites in the heart, brain and kidney

Organ	Relative distribution
Heart	atria > ventricle
Kidney	glomeruli > inner medulla > cortex
Brain	cerebellum > hypothalamus > striatum
Adrenal gland	medulla > cortex

including a relatively high concentration of sites in the lungs (Neuser et al. 1989), in the renal cortex (Kohzuki et al. 1989) and the cerebellum (Davenport et al. 1989b). Binding sites were even found to be present in the endocardial cells of the myocardium and the cardiac valve cusps (Power et al. 1989a, b), in nerve trunks, in the adventia of blood vessels, in the gray matter of the spinal cord and even in tumours associated with the adrenal gland (Conn's adenoma and phaeochromocytoma). However, as can been seen from Table 7.1, endothelin binding sites are not found in all tissues. They are absent, for example, from cartilage, connective tissue, red blood cells and platelets. In addition (Table 7.2) their distribution in certain organs – including the heart, kidneys and brain – lacks homogeneity.

Characterization of Endothelin Binding Sites

Competition and saturation binding studies using isolated membrane fragments harvested from a variety of tissues – including the myocardium (Gu et al. 1989a, b;

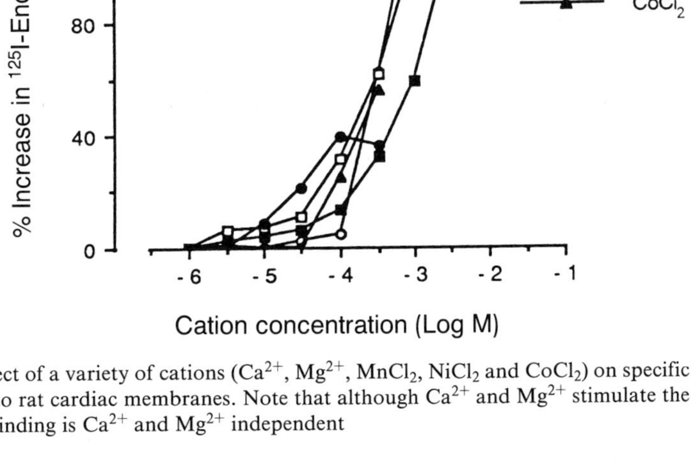

Fig. 7.2. Stimulant effect of a variety of cations (Ca^{2+}, Mg^{2+}, $MnCl_2$, $NiCl_2$ and $CoCl_2$) on specific endothelin-1 binding to rat cardiac membranes. Note that although Ca^{2+} and Mg^{2+} stimulate the specific binding, the binding is Ca^{2+} and Mg^{2+} independent

Nayler 1989; Nayler et al. 1989a, b), neonatal cardiocytes (Hirata et al. 1988, 1989a; Hirata 1989; Galron et al. 1989), mesangial cells (Sugiura et al. 1989), placenta (Nakajo et al. 1989), uterus (Bousso-Mittler et al. 1989), lung (Masuda et al. 1989) and cerebellum (Kloog et al. 1989) – as well as radioautographical studies on tissue slices (Chester et al. 1989; Turner et al. 1989a, b; Kohzuki et al. 1989; Davenport et al. 1989a, b) have provided a wealth of information relating to the properties of these binding sites. These properties can be summarized as follows:

(i) binding is saturable, and temperature sensitive (Gu et al. 1989c);
(ii) specific binding accounts for almost 80 percent of the total binding;
(iii) binding occurs in the absence of added Ca^{2+} or Mg^{2+} but is stimulated by a variety of divalent cations, including Ca^{2+} and Mg^{2+} (Figure 7.2, and Gu et al. 1989b, c);

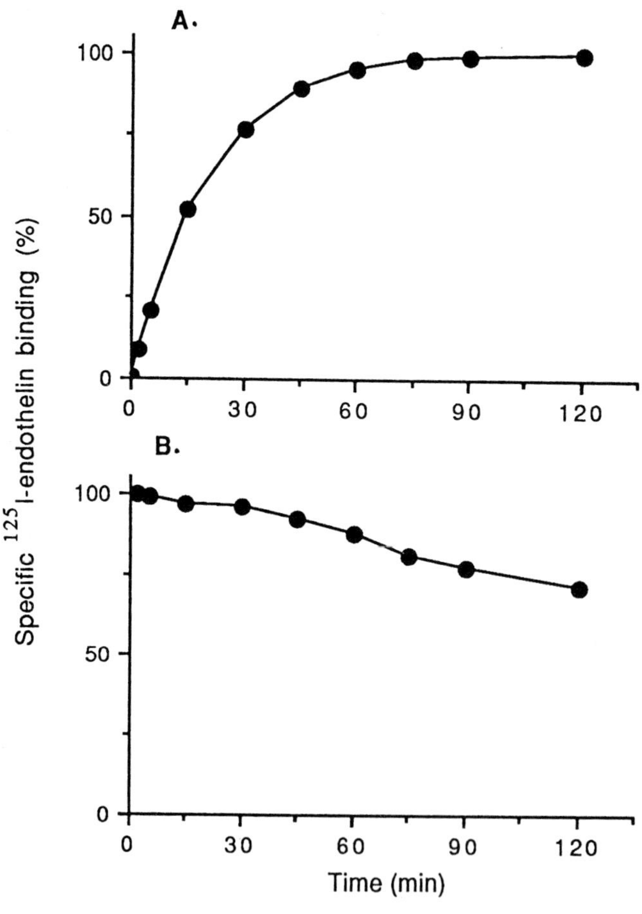

Fig. 7.3. Association (A) and dissociation (B) kinetics of [^{125}I]endothelin-1 binding to rat cardiac membranes. Ordinate scale: specific [^{125}I]endothelin binding as %, taking binding at 120 minutes as 100%. (B) is the dissociation curve, calculated relative to 100% association at 120 minutes.

Table 7.3. Effect of a variety of agents, including calcium antagonists and agonists, on the specific binding of [^{125}I]endothelin to rat ventricular membranes

Agents	Percentage inhibition of binding
Sarafotoxin	100
Endothelin	100
Calcium Antagonists	
(i) L-channels	
Nicardipine	0
Nifedipine	0
d-cis Diltiazem	0
(–) D888	0
Isradipine	0
Flunarizine	0
Lidoflazine	0
Cobalt chloride	0
(ii) T-channels	
Nickel chloride	0
(iii) N-channels	
w conotoxin	0
Calcium Agonists	
Bay K8644	0
Na^+:H^+ Exchange Inhibitor	
Amiloride	0
α_1 Adrenoceptor Antagonist	
Prazosin	0
Vasoactive Polypeptides	
Vasopressin	0
Angiotensin	0
β-adrenoceptor Agonist	
Prenylamine	0
Coronary Dilator	
Glyceryl trinitrate	0
Coronary Constrictor	
Ergonovine	0

These binding studies were performed at 37°C and a wide range of concentrations were used. The details were given in reference Gu et al. 1989a, b, c

(iv) binding is relatively slow, 60 minutes or more being required to reach asymptote during incubation at 37°C (Figure 7.3). Dissociation is also slow. These slow rates of association and dissociation almost certainly account for endothelins persistent biological activity;

(v) the binding is insensitive to a wide variety of substances, including (Table 7.3, and Figure 7.4) inhibitors of the L-, N- and T-type Ca^{2+} channels, and other vasoactive polypeptides – including angiotensin II and vasopressin. It is also insensitive to inhibitors of the Na^+:H^+ exchanger, and to α_1 adrenoceptor antagonists (prazosin) and β_1 agonists (isoprenaline);

(vi) monoclonal antibodies raised against the dihydropyridine sensitive Ca^{2+} channels do not recognize the endothelin binding sites (Miyazaki et al. 1990);

(vii) the solubilized endothelin-1 binding site has a sedimentation coefficient of 135 whereas the solubilized dihydropyridine binding site has a sedimentation coefficient of 225 (Miyazaki et al. 1990).

(viii) unlabelled endothelin-1 and sarafotoxin S6b competitively inhibit endothelin binding (as can be seen from Figure 7.4, and Gu et al. 1989a, c; Kloog et al. 1989);

(ix) the binding is pH sensitive, but the optimal pH is tissue specific. In aorta (Kanse et al. 1989) and heart (Gu et al. 1989a, c), for example a pH of 7.0 is optimal whereas in lungs the optimal pH is around 3.0 (Kanse et al. 1989);

(x) the binding is not inhibited by the neuropeptide NPY, or by the chemically homologous bee venom, apamin (Davenport et al. 1989a, b). In addition, (Yamasaki et al. 1989) neither the atrial natriuretic peptide nor inhibitors of the Na^+ channel (tetrodoxin and saxitoxin) inhibit the binding.

However

(xi) the binding is endothelium independent;

(xii) it is destroyed by Triton X-100 and the sulphydril reducing agent dithiothreitol (Kanse et al. 1989);

(xiii) in human epicardial coronary arteries it is unaffected by atherosclerosis or intimal proliferation (Power et al. 1989b) but increases during myocardial ischaemia (Chester et al. 1989);

(xiv) the binding sites have been solubilized (Miyazaki et al. 1990; Liu et al. 1990b), without loss of activity, and

(xv) the affinity of the binding sites closely approximates the concentration required to elicit a physiological (vasoconstriction or positive inotropy) or biochemical (raised cytosolic Ca^{2+} and enhanced phosphoinositol metabolism) response.

In general, therefore, the distribution of the endothelin binding sites is widespread. They are specific for endothelin and do not coincide or even co-distribute with any of the known Na^+ or Ca^{2+} channels. However it is still theoretically possible (but unlikely) that the polypeptide interacts with hitherto unrecognized sites in these channels – particularly the Ca^{2+} channels.

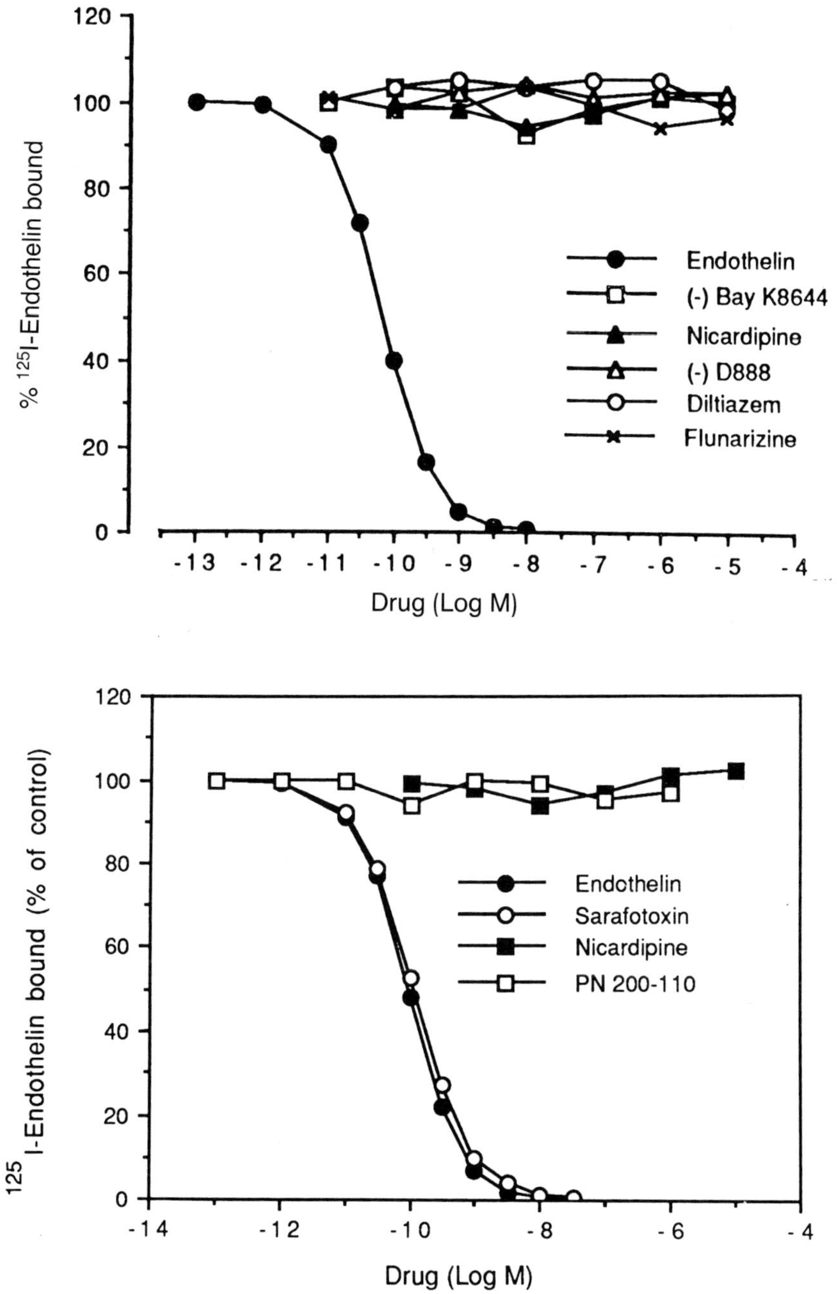

Fig. 7.4 A, B. Effect of Bay K8644 (calcium agonist) and a variety of calcium antagonists and sarafotoxin (SRTX-b) on the specific binding of [^{125}I]endothelin-1 to isolated cardiac membranes from normotensive rats. Note that neither the Ca^{2+} agonist nor the calcium antagonists competed with endothelin-1 for its specific binding site. Sarafotoxin SRTX-b (Chapter 4) however, competitively inhibited the binding, confirming that it interacts with the same binding site

Evidence of Multiple Endothelin Receptors

It now seems probable that there are sub-populations of endothelin receptors, some preferring endothelin-1 and -2, and the others preferring endothelin-3. This evidence comes from two sources – one involving the biochemical analysis of the receptors, and the second involving the pharmacological consequences of their stimulation.

The Biochemical Analysis of the Endothelin Receptors

The sub unit and amino acid composition of the endothelin receptor complex has yet to be established. However a start has been made in that the receptor complex has been solubilized without loss of affinity or selectivity (Ambar et al. 1989; Nakajo et al. 1989; Miyazaki et al. 1990; Suguira et al. 1989c; Liu et al. 1990b). Affinity labelling of the solubilized complex appears to have provided firm evidence for the existence of multiple receptors in some tissues. For example (Table 7.4) two types of receptors have been identified in chick cardiac membranes – one with a molecular weight of around 50 KDa which preferentially binds endothelin-1, and a second smaller complex (molecular weight of 46 or 34 KDa) which preferentially binds endothelin-3. A similar situation seems to occur in rat lung membranes, where two receptors have been identified – an endothelin-1 preferring receptor with a molecular weight of 44 KDa, and an endothelin-3 preferring receptor with the slightly smaller molecular weight of 32 KDa (Sugiura et al. 1989c). Multiple receptors may not be present in all tissues, however, because affinity labelling of solubilized human placenta membranes only revealed the presence of a single endothelin binding protein. This had a molecular weight of 32 KDa. However, at this stage it is important to appreciate that these preliminary biochemical studies

Table 7.4. Biochemical evidence favouring the existence of multiple endothelin receptors

Tissue	Receptor M.W. (KDa)	Selectivity	Reference
Rat lung	44	ET-1 > ET-3	Masuda et al. 1989
	32	ET-3 > ET-1	
Chick heart	53	ET-1 > ET-3 > SRTX	Watanabe et al. 1989
	46	ET-3 > ET-1 > SRTX	Watanabe et al. 1989
	34	ET-3 > ET-1 > SRTX	Watanabe et al. 1989
Human placenta	32	ET-1 > ET-3	Nakajo et al. 1989
Rat mesangial cells	58	ET-1 > ET-3	Sugiura et al. 1989c
	34	ET-1 > ET-3	

ET-1 = endothelin-1, ET-3 = endothelin-3, SRTX = sarafotoxin S6b

Note that the molecular weights of the endothelin receptor complex is far smaller than that of the dihydropyridine binding site in the L-type Ca^{2+} channel, which is around 190 KDa really only confirm the presence of multiple endothelin receptors in some tissues, and although these receptors *may* subserve different endothelin-dependent functions (one set perhaps motivating the events which culminate in Ca^{2+} influx and the other being concerned with the phosphoinositol cascade) evidence supporting this idea is still in its preliminary form. This preliminary data, however, is supported by the apparent tissue dependent differences in the concentration of endothelin needed to occupy fifty percent of the binding sites. For example in rat ventricular membranes the IC_{50} value (the concentration needed to occupy fifty percent of the binding sites), shows the following rank order ET-2 $<$ ET-1 $<$ ET-3. The lung membranes exhibit the same rank order, but with ET-2 $<<$ ET-1 $<<<$ ET-3. In the brain the order is different: ET-2 $<$ ET-1 $=$ ET-3. This means that the ET receptors in the lung and ventricular membranes have a higher affinity for ET-1 than ET-3, whereas in the brain the receptors display an equal affinity for ET-1 and ET-3.

Are the Endothelin Binding Sites Functional?

This question can only be answered in the affirmative, because chemical manipulations of the peptide show parallel changes in binding affinity, physiological response and activation of the phosphoinositol pathway (Hirata et al. 1989e, 1990; Maggi et al. 1989b, c, 1990; Nakajima et al. 1989; Randall et al. 1989). The manipulations which have been made so far have included removing the terminal tryptophan residue, removing the terminal hexapeptide, and disrupting both disulphide bridges between the cysteine residues.

"Up" and "down" Regulation of the Endothelin Receptors

Most membrane-located receptors are capable of being "up" and "down" regulated. For example, β_2 adrenoceptors in peripheral sympathetic ganglia of spontaneously hypertensive rats are "up" regulated – meaning that their density is increased. Endothelin-1 receptors are no exception. As far as the heart is concerned they "up" regulate in response to cyclosporine administration (Nayler et al. 1989b), during prolonged ischaemic episodes (Chapter 14), and after exposure to free radicals (Nayler, unpublished data). They also "down" regulate in response to streptozotocin-induced diabetes (Nayler et al. 1989c), after exposure to endothelin (Hirata et al. 1988) and possibly in some tissues including the atria, in response to systemic hypertension (Gu et al. 1990). Such an effect is evident in the data given in Table 7.5, which shows that the density of binding sites in ventricular membranes prepared from normotensive rats $>$ that obtained for age-matched hypertensives (Gu et al. 1990). Under these same conditions the density of the receptors in the brain seems to have "down" regulated in response to the development of spontaneous hypertension. As far as rat vascular smooth muscle cells are concerned many factors can trigger "down" regulation – histamine, serotonin, angiotensin II, atrionatriuretic peptide, and phorbol -12, 13 dibutyrate. Moreover, as Figure 7.5 shows, the effect is dose-dependent, at least for the effect of angiotensin II and phorbol -12, 13 dibutyrate.

Table 7.5. Effect of spontaneous hypertension on the density (B_{max}) of endothelin-1 binding sites (identified with [^{125}I]endothelin-1) in rat ventricular and brain membranes

Endothelin binding site density (B_{max}) (fmol/mg protein)		
Membrane Source	Normotensive	Hypertensive
Left ventricle	89.3 ± 5.0	53.7 ± 8.2 ($p < 0.02$)
Atria	114.7 ± 9.4	37.0 ± 12.0 ($p < 0.01$)
Brain	129.5 ± 5.0	167.2 ± 9.6 ($p < 0.01$)

Note that whereas the density (B_{max}) of the cardiac receptors declined, the density in the brain increases. The rats were age matched (Ref. Gu et al. 1990)

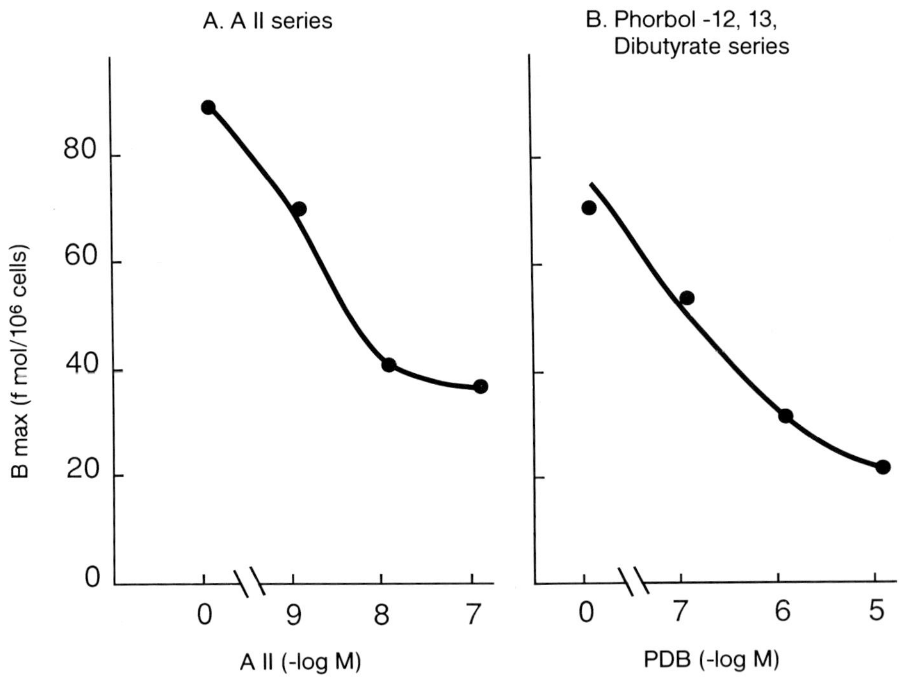

Fig. 7.5. Down regulation of specific endothelin-1 binding sites by angiotensin II and phorbol -12, 13 dibutyrate. Note that it is the density (Bmax) of binding sites which was being measured here, using isolated porcine aortic smooth muscle cells which had been exposed for 15 hours to the angiotensin II and phorbol ester. Binding site density was measured in the absence of these compounds. (Data from Roubert et al. 1989)

Table 7.6. The relative density of endothelin-1 binding sites assayed for isolated membranes, whole cells and tissue slices

Tissue Isolated Membrane Studies	ET-1 density of binding (fmol/mg protein)
Rat cardiac membranes	95
Rat lung membranes	1,650
Rat uterus membranes	220
Rat renal glomeruli	1,500
Human placenta	240
Porcine aortic membranes	1,250
Isolated Cells	
Rat aorta	10,000–13,000 sites/cell
Retinal capillary pericytes	100,000 sites/cell
Rat neonatal cardiocytes	8,000–130,000 /cell
Human fibroblasts	156/cell
Radioautography on Tissue Slices	
Guinea pig heart	207 fmol/mg tissue
Guinea pig lung	1,215 fmol/mg tissue
Human hypothalamus	625 fmol/mg protein
Human cerebellum	776 fmol/mg protein

The Density of Endothelin-1 Binding Sites in Various Tissues

Although Table 7.4 provided some data relative to the distribution of the endothelin binding sites in different tissues, it did not provide data concerning absolute densities of binding. This is shown in Table 7.6 – using data obtained from several different techniques, including isolated membrane studies, binding to isolated cells, and binding identified by radioautography using tissue slices. The data has mainly been obtained from the literature. However it confirms the fact that some tissues contain more endothelin-1 binding sites than others. The lungs, for example contain an abundance of binding sites, as do the renal glomeruli.

Summary

1. Specific membrane-located endothelin binding sites occur in a wide variety of tissues. They are not restricted to the vasculature.
2. These binding sites are functional and subject to "up" and "down" regulation.
3. Various pathological conditions, including hypertension, ischaemia (see Chapter 14), cyclosporine-induced toxicity and streptozotocin-induced diabetes modifies the density of the binding (Table 7.7).
4. Binding of endothelin to its receptor(s) is slow, pH sensitive but Ca^{2+}-independent.

Table 7.7. Conditions which "up" and "down" regulate the [^{125}I] endothelin-1 binding site density

	Tissue	Reference
"Up" Regulation		
Ischaemia	Cardiac	Liu et al. 1990a
Post ischaemic reperfusion	Cardiac	Liu et al. 1990a
Cyclosporine toxicity	Cardiac	Nayler et al. 1989b
Cholera toxin	Vascular Smooth Muscle	Roubert et al. 1990
Hypertension	Brain	Gu et al. 1990
"Down" Regulation		
Streptozotocin-induced diabetes	Cardiac	Nayler et al. 1989c
Hypertension	Cardiac	Gu et al. 1990
Endothelin	Smooth muscle cells	Hirata et al. 1988

5. The endothelin receptors discriminate between other vasoactive peptides, and calcium and sodium channel antagonists. Presumably therefore, they are a unique class of binding sites.
6. The sarafotoxins, the chemically homologous toxins obtained from the venom of the asp (Chapter 4), interact with the endothelin binding sites.
7. The endothelin receptor binding sites can probably be subdivided into several subpopulations: one with a molecular weight of around 50 KDa and the other, around 35 KDa. These receptors may subserve different functions.

Chapter 8

Metabolism and Clearance of the Endothelins

"Surely you have noticed... but no, why should you. Noticing things is only a trouble."
Toad's advice to The Rat, in "Wind in the Willows", by KENNETH GREENE.

Because endothelin-1 is such a potent vasoconstrictor, reports of its presence in circulating blood surprised many investigators. This element of surprise was unnecessary, however, because, as outlined in Chapter 10, the plasma levels are usually quite low. Even so, the role of the circulating endothelin-1 is unknown. Possibilities include:
(i) acting as a circulating hormone, and
(ii) acting as a local hormone, with some secondary leakage into the circulation.

Irrespective of whether either of these conditions apply, the plasma levels of endothelin-1 are normally well regulated and maintained at a relatively low level usually around 1 pg per ml (Chapters 9 and 10). This is well below the threshold level for vasoconstriction.

From a purely theoretical point of view regulation of the endothelin level could involve:
(i) a regulated rate of production and secretion, and/or
(ii) an efficient mechanism of removal.

This chapter is primarily concerned with the second of these two possibilities – that is, the mechanisms involved in removing endothelin-1 (and presumably the other isoforms) from the circulation.

Time Course of Endothelin-1 Removal

Experiments in which radioactively labelled endothelin-1 has been injected either intraarterially or intravenously into anaesthetized rats (Anggard et al. 1989; Shiba et al. 1989) and guinea pigs (Pernow et al. 1989b) have provided some valuable information concerning its rate of removal from the circulation. The disappearance curves (Figure 8.1) show that:
(i) the removal pattern fits a two compartment exponential model, with
(ii) more than sixty percent of the endothelin-1 disappearing within the first few minutes (Figure 8.1),

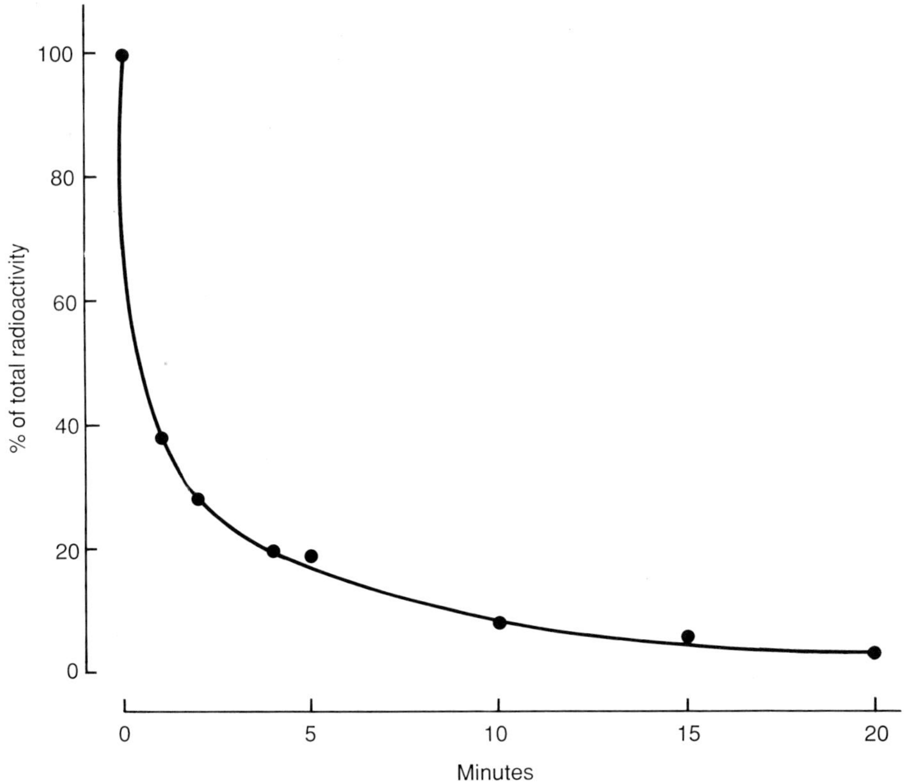

Fig. 8.1. Time-course for the disappearance of iodinated endothelin-1 from the blood stream of an anaesthetized rat. The iodinated endothelin-1 was injected intravenously

(iii) and the remainder being removed at a relatively slow rate, with a half-life of around forty-five minutes.

There are other points which are worth noting, including the fact that

(iv) the disappearance of endothelin-1 from the circulation does *not* seem to be accompanied by the appearance of metabolites, and

(v) the radioactively-labelled peptide accumulates in certain tissues, including the lungs and kidneys, and to a lesser extent, the liver, spleen and heart (Shiba et al. 1989).

Hence, the curve describing the removal of endothelin-1 from the circulation fits a two compartment model, with an early rapid phase followed by a much slower phase. More than sixty percent of the endothelin is removed during the rapid phase, which lasts for about two minutes. The disappearance of the polypeptide is accompanied by its accumulation in other tissues, including the kidneys and lungs, but apparently without the formation of metabolites.

Table 8.1. Removal of polypeptides from the circulation

Polypeptide	Half-life	Method of removal
Bradykinin	17 seconds	Inactivation by circulating carboxypeptidases
Angiotensin II	2–4 minutes	Removed by kidneys, liver and hindquarters, plus inactivation by circulating aminopeptidases
Atrial natriuretic peptide	<60 seconds	Binding to "clearance" receptors in the vasculature
Endothelin-1	1.5–7 minutes	? Binding to receptors in kidneys, lung and liver

Comparison with Other Vasoactive Polypeptides

Relative to the rate at which other vasoactive polypeptides are removed from the circulation, the rate of removal of endothelin-1, (and endothelin-3) (Anggard et al. 1989), is slow. *Bradykinin*, for example, has a half-life of less than twenty seconds. Its removal involves its inactivation by a carboxypeptide (kinase I). *Angiotensin II* (Table 8.1) has a slightly longer half-life, of between one and two hundred seconds. It's removal from the blood stream involves several tissues, including the kidneys and liver. It is inactivated by an aminopeptidase (Table 8.1). Another peptide, the *atrionatriuretic peptide*, is removed from the circulation at a very rapid rate. It has a half-life of less than thirty seconds, and its removal involves its binding to high affinity sites in the vasculature, these sites then serving as "clearance receptors". In general, therefore, relative to other vasoactive peptides, the rate at which endothelin-1 is removed from the circulation is not so very fast.

Comparatively little information is available concerning the relative rates of removal of the individual endothelin isoforms, but endothelin-1 is removed more rapidly than endothelin-3 (Anggard et al. 1989).

Mechanisms Involved in the Removal of Endothelin-1

Since the endothelins are polypeptides it might have been expected that their removal from the circulation would involve degradation by circulating peptides. This idea is unacceptable, because:

(i) endothelin-1 is stable in blood at 37 °C for at least half an hour (Pernow et al. 1989b); and
(ii) degradation products do not appear in the circulation.

Another possibility which can be discarded is that the endothelins are degraded by the angiotensin converting enzyme. This idea is also untenable, because the rate of disappearance of endothelin-1 from the circulation is unaffected by captopril – a potent inhibitor of this enzyme (de Nucci et al. 1988).

The most likely explanation for the relatively rapid disappearance of the endothelins is that the removal process involves *receptor binding followed by internalization*. This hypothesis is supported by the following facts:

(i) when injected into the circulation, radioactively labelled endothelin-1 accumulates preferentially in certain tissues, particularly the lungs and kidneys (Pernow et al. 1989b; Anggard et al. 1989);

(ii) this process of "accumulation" is rapid, with sixty percent of the polypeptide leaving the circulation during one passage through the organ; and

(iii) those organs, including the lungs and kidneys, which preferentially accumulate the endothelins have particularly high endothelin binding site densities (Chapter 7).

It is not known, as yet, whether the receptor binding is followed by internalization and subsequent metabolic degradation. In the kidneys, however, a large proportion of the endothelin must be eliminated without degradation, because urine normally contains around 30 pg endothelin-1 per ml (Berbinschi and Ketelslegers 1989). This means that the concentration in urine is around fifteen times greater than that of plasma.

Dissociation Between the Disappearance of Endothelin-1 from the Bloodstream and it's Constrictor Effect

One of the interesting properties of endothelin-1 is that although its injection into the blood stream causes a significant and sustained pressor response, the response develops relatively slowly. Thus, the maximum hypertensive response requires about fifteen minutes to develop, by which time almost all of the injected endothelin-1 must have disappeared from the circulation. There are several ways of explaining this phenomenon:

(i) the sustained pressor response may be due to the prolonged binding of the polypeptide to its receptors (Chapter 7), or

(ii) activation of the receptors may trigger a cascade of events which results in a prolonged activation of the vascular smooth muscle contractile proteins.

In either case, plasma concentrations of endothelin-1 cannot be expected to correlate with systolic blood pressure – even though they may be raised in certain hypertensive states.

Summary

The disappearance of endothelin-1 from the circulation involves two phases – a rapid phase, which is complete within one or two minutes and accounts for the removal of over sixty percent of the polypeptide, and a much slower phase, with a

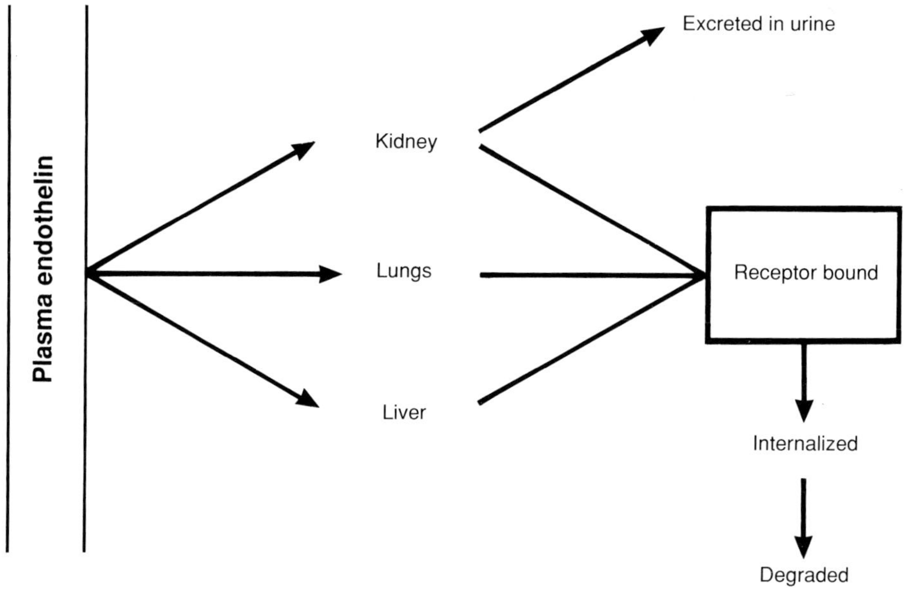

Fig. 8.2. Schematic representation of events which may be responsible for the disappearance of endothelin-1 from the blood stream

half life of around forty five minutes. Endothelin moves from the circulation to accumulate in certain tissues (Figure 8.2), mainly the lungs and kidneys – tissues which are richly endowed with endothelin binding sites. Whether receptor binding is followed by internalization and metabolic degradation has yet to be established. However large amounts of endothelin-1 are excreted unchanged in urine.

The rapid time-course for the disappearance of endothelin-1 from the blood stream contrasts sharply with the slow development and sustained nature of its pressor effect (Yanagisawa et al. 1988a; Clarke et al. 1989).

Chapter 9

Plasma, Tissue and Urine Endothelin-1

> *"What size would you like to be?"*
> *"Oh, I'm not particular as to size", Alice hastely replied, "only one doesn't like changing so often, you know".*
> Advice from a Caterpillar: LEWIS CAROL, in: Alice in Wonderland.

When Yanagisawa and his colleagues began investigating the processes which control the synthesis of endothelin-1, they started with porcine cultured aortic endothelial cells because it was in the supernatant of these cells that the polypeptide was first detected (Chapter 1). However, once reliable and specific immunoreactive techniques were developed (Saito et al. 1989a, b; Ando et al. 1989) it became apparent that neither the distribution nor the synthesis of the endothelins is limited to vascular endothelial cells. To the contrary, these peptides are synthesized in many other tissues – including the brain and spinal cord, the lungs, kidneys, heart, spleen, duodenum, trachea, and even in tumours. In addition it soon became known that:

(i) endothelin-1 is a normal constituent of plasma, urine, lymph, cerebrospinal fluid, human milk and bronchial exudates; that
(ii) its receptors are not limited to the cardiovascular system; and that
(iii) the plasma, lymphatic, urinary and cerebrospinal levels vary under certain physiological and pathological conditions.

This chapter describes the tissue distribution of endothelin-1, its concentration in plasma, urine and human milk and the factors which affect its rate of synthesis, secretion and removal. At the outset several important points emerge:

(i) as far as tissue concentrations are concerned, the concentration of endothelin-1 in the inner medulla of the kidney $>>$ lung $>>$ heart (atrium and ventricle) $>$ brain $>$ duodenum;
(ii) this rank order of concentration is *not* species specific;
(iii) as far as the brain is concerned endothelin-1 accumulates in specific areas, including the hypothalamus, and is synthesized in the posterior pituitary system (Yoshizawa et al. 1990 and Chapter 12);
(iv) the polypeptide has a short half-life in circulating blood, and seems to be removed through the kidneys and the lungs;
(v) conditions which are associated with tissue injury – including thrombin formation - accelerate its rate of synthesis and secretion; and
(vi) the endothelial derived nitric oxide component of the endothelial-derived relaxing factor (EDRF) inhibits the thrombin-induced stimulation of endothelin-1 production (Boulanger and Lüscher 1990).

The corollary to this inhibitory effect of nitric oxide is that if a thrombus forms under conditions in which the nitric oxide producing capacity of the endothelium is impaired – as occurs during conditions of hyperlipidaemia, hypertension and atherosclerosis, and when the endothelium is regenerating after injury – the thrombin-induced stimulation of endothelin production will continue unopposed resulting in the possible triggering of vasospasm and smooth muscle cell proliferation. Such a sequence of events may easily be involved in the generation of an atherosclerotic lesion, as discussed in Chapter 13.

Tissue Levels of Endothelin-1

Kidney, Lung, Heart and Brain

Most of the information which is currently available concerning the tissue levels of endothelin has come from the laboratories of Kitamura (Kitamura et al. 1989 a,b), Shinmi (Shinmi et al. 1989a, b), Pernow (Pernow et al. 1989b) and Yoshimi et al. 1989. Tissues from rats and pigs have usually been used (Table 9.1), with the consistent finding of the concentration in the inner medulla of the kidney >> lung >> heart > brain > duodenum. Overall, however, the tissue levels are low – so low,

Table 9.1. Tissue levels of endothelin-1

Tissue	Concentrations (pg per mg wet wt)	
	Porcine	Rat
Brain cortex	0.58 ± 0.08[a]	0.65 ± 0.09[b]
Hypothalamus	0.69 ± 0.09[a]	0.72 ± 0.09[a]
Lung	2.19 ± 0.09[a]	3.41 ± 0.82[a]
	1.5 ± 0.1[d]	
Heart		
atrium	0.94 ± 0.08[a]	
ventricle	0.91 ± 0.09[a]	
Liver	0.25 ± 0.36[a]	0.22 ± 0.03[a]
Spleen	0.57 ± 0.10[a]	0.97 ± 0.15[a]
	0.28 ± 0.04[d]	
Kidney		
cortex	0.36 ± 0.05[a]	0.58 ± 0.07[a]
	0.18 ± 0.01[d]	
outer medulla	6.4 ± 1.1[a]	8.7 ± 2.2[a]
Adrenal gland	0.24 ± 0.04[a]	0.16 ± 2.2[b]
Duodenum	0.34 ± 0.04[a]	0.39 ± 0.04[a]
Spinal cord	0.12[c]	

[a] Kitamura et al. 1989a
[b] Kitamura et al. 1989b
[c] Shinmi et al. 1989a
[d] Pernow et al. 1989b

in fact, that some investigators question their physiological significance. For example, the concentration in porcine heart muscle is in the region of 0.9 pg per mg wet weight (Table 9.1), a concentration which is well below that needed to evoke a positive inotropic response (Chapter 5). There are several possible explanations for this apparent discrepancy between the concentration of endothelin which is needed to evoke a physiological response and the tissue levels.

(i) It may be the turn-over rate rather than the absolute concentration of endothelin-1 which matters;
(ii) measurements of total tissue content will provide little information concerning localized areas of high endothelin concentration; and
(iii) under pathological conditions the synthesis of endothelin-1 may be up- or down-regulated, depending on the particular pathology which is involved.

The next question to consider is why both the inner medulla of the kidney and the lungs contain unusually high levels of endothelin-1? In the inner medulla of the rat kidney, for example, concentrations of 8.7 ± 2.2 pg per mg wet tissue weight have been described (Kitamura et al. 1989b). Porcine kidney inner medulla contains similar amounts (Table 9.1). The concentration in rat (3.4 ± 0.82 pg per mg wet tissue weight) and porcine (2.19 ± 0.09 pg per mg wet tissue weight) lungs is certainly less than that of the kidney inner medulla but even so it is well in excess of that found in many other tissues (Table 9.1).

As far as the inner medulla of the kidney is concerned the high levels of endothelin-1 are matched by an abundance of receptors (Kohzuki et al. 1989). It is possible, therefore, that endothelin may regulate urine formation, as well as being a vasoconstrictor. The lungs also contain an abundance of endothelin receptors and here it has been suggested, but not yet unequivocally proved, that endothelin is taken up for degradation (de Nucci et al. 1988). Not all investigators share this view, however. Pernow, for example, has recently concluded that endothelin is cleared from the arterial circulation through the kidneys (Pernow et al. 1989b).

Certain regions of the brain – including the paraventricular and supraoptic nuclear neurons in the *hypothalamus* and the neurohypophysis (posterior pituitary) (Yoshizawa et al. 1990) - contain relatively large amounts of endothelin-1, and since endothelin-1 mRNA is present in the paraventricular nuclear neurons (Yoshizawa et al. 1990) it must be assumed that the polypeptide is produced there. It is doubtful whether any of the endothelin-1 which is produced in the posterior pituitary system contributes to the circulating level of the polypeptide, because when injected, endothelin-1 does not cross the blood brain barrier. A more likely possibility is that when released it modulates the release of the neurosecretory hormones – vasopressin and oxytocin. As far as vasopressin is concerned, the intravenous infusion of endothelin-1 has been shown to promote vasopressin release (Goetz et al. 1988).

Plasma

Early attempts at measuring plasma levels of endothelin-1 gave falsely high results, probably because investigators failed to realize that serum albumin interferes with

Table 9.2. Plasma and urine levels of endothelin-1 in healthy volunteers

	Reference
Plasma level (pg per ml)	
0.3–1.7	Hartter and Woloszczok 1989
1.5	Miyauchi et al. 1989
1.4	Ando et al. 1989
1.5	Saito et al. 1989a, b
0.3–0.69	Cernacek et al. 1989
Urine level (pg per ml)	
33.1	Berbinschi and Ketelslegers 1989
Milk level (human) pg per ml	
160	Lam et al. 1990

Values are either mean or mean ± SEM. The plasma levels of endothelin-1 are unaffected by gender. Saito et al. (1989) detected levels of 1.5 ± 0.04 pg per ml in healthy volunteer males, compared with levels of 1.5 ± 0.5 pg per ml in females. The plasma levels do change, however, under certain pathological conditions, including uraemia, cardiac infarction, surgical stress and cardiogenic shock, hypertension and after kidney transplantation (Chapter 10)

the radioimmune assay for endothelin. This difficulty can be avoided, however, by passing the plasma through a Seppak column (Cernacek and Stewart 1989). When this precaution has been taken, plasma levels of around 1.0 pg per ml have been obtained for healthy volunteers (Cernacek and Stewart 1989). Some of the values which have been obtained are listed in Table 9.2. As with the tissue levels, the low plasma levels raise questions as to its physiological significance. However, plasma levels may be misleading because:

(i) if endothelin is released not only into the lumen of the vessel, but also in the abluminal direction, relatively high concentrations may be achieved at the vessel wall close to its site of production; and

(ii) the rapid clearance of endothelin from the circulation (with a half-life of around eighty seconds) could mean that a localized accumulation of endothelin would be missed. This rapid clearance from the circulation is in no way at odds with the sustained constrictor effect of the polypeptide (Chapter 5), a property which can be accounted for simply in terms of its persistent binding to its receptors (Chapter 7).

Urine

Surprisingly little attention has been given so far to the measurement of urinary levels of endothelin-1. However, immunoreactive endothelin-1 is present in urine (Berbinschi and Ketelslegers 1989), and the levels are well in excess of the plasma levels. There are no sex related differences, and in healthy volunteers (mean age 25 years) levels of about 33 pg per ml have been obtained, with a daily excretion rate of around 40 ng per 24 hours (41.5 ± 17 ng per 24 hours in 37 healthy volunteers). These relatively high urinary levels (Table 9.2):

(i) support the concept of clearance through the kidneys as being a major route of clearance; and
(ii) point towards a relatively high turnover rate of endothelin-1.

Bronchial Exudates

Here the levels are even lower than those found in plasma. Normally they seem to be around 0.05 pg per ml of lavage fluid (Nomura et al. 1989).

Cerebrospinal Fluid

Levels ranging between 3.5 and 12 pg per ml have been obtained for healthy volunteers (Hoffman et al. 1989). This is about seven times *higher* than the plasma level.

Pulmonary Lymph

Although attempts at detecting endothelin-1 in pulmonary lymph have failed – presumably because the levels are below the limits detectable by the currently available assay methods, levels in excess of 35 pmol/litre have been detected during endotoxin infusion (Morel et al. 1989; Pernow et al. 1989a). Maybe this is just a "spill-over" effect from the accompanying increase in plasma endothelin-1 which occurs under these conditions.

Human Milk

Milk from lactating mothers contains as much as 160 pg endothelin-1 per ml (Lam et al. 1990), a level which is well in excess of that found in plasma or urine. The presence of such a high level raises questions relating to its source and possible function. Since the concentration greatly exceeds that found in plasma (Table 9.2) it follows that either the mammary gland must possess an active concentrating mechanism or the endothelin must be of extravascular origin. As for its function, since the gastrointestinal tract contains endothelin receptors and since the polypeptide causes the gut to contract, the presence of endothelin in the milk of lactating mothers may, as Lam et al. (1990) have suggested, be involved in the development of motility in the gastrointestinal tract of the neonate.

Sites of Endothelin Synthesis

The techniques used to identify sites of endothelin synthesis include:
(i) measurement of endothelin mRNA; and
(ii) immunohistochemical localization of immunoreactive endothelin-1.

The synthesis of endothelin-1 is certainly not limited to vascular endothelial cells, although it's rate of production in these cells exceeds that at any other known site (Table 9.3). Other sites of production include the kidneys (Kosaka et al. 1989), the spinal cord (Shinmi et al. 1989a), the human umbilical vein and placenta (Itoh et al. 1988), certain tumour cells (Suzuki et al. 1989) and the posterior pituitary system (Yoshizawa et al. 1990). The aortic endothelial cells, however, seem to provide the fastest rate of production, at least in culture.

The presence of endothelin and its precursors in the spinal cord is of interest, because it seems to be concentrated in the motorneurones, the dorsal horn neurones and the dot and fibre-like structures in the dorsal horn of the lumbar spinal cord. Maybe endothelin does act as a neuropeptide within the central nervous system.

The significance of the unexpectedly widespread distribution of endothelin-producing sites is unknown. Possibly, as Yanagisawa and Masaki (1989a, b) have

Table 9.3. Sites of Endothelin-1 synthesis

Tissue		Mean rate of endothelin-1 production (pg endothelin per 10^6 cells per 2 days)	Reference
Cultured porcine aortic endothelial cells		14,630	Kosaka et al. 1989
Cultured kidney cells			
Species	Cell line		
Monkey	COS-7	1,310	Kosaka et al. 1989
	CV-1	2,460	Kosaka et al. 1989
	Vero	554	Kosaka et al. 1989
Rabbit	RK-13	240	Kosaka et al. 1989
Rat	NRK-52E	530	Kosaka et al. 1989
Dog	MDCK	1,115	Kosaka et al. 1989
Hamster	BHK-21	110	Kosaka et al. 1989
Brain		unknown	Kosaka et al. 1989
Lung		unknown	Kosaka et al. 1989
Spinal cord		unknown	Shinmi et al. 1989a
		unknown	Giaid et al. 1989
Human umbilical vein		unknown	Itoh et al. 1988
Human placenta		unknown	Itoh et al. 1988
Tracheal epithelial cells		unknown	Black et al. 1989
Tumour cells		unknown	Suzuki et al. 1989

suggested, the localized production of endothelin may be indicative of its importance as a *local* regulator of cell function. Alternatively the capacity to produce significant amounts of endothelin locally may be of considerable importance only under pathophysiological or emergency conditions, where haemostasis and wound repair become vital for survival. As a corollary, disturbances in or breakdown of the mechanisms which regulate endothelin synthesis and its secretion could easily precipitate, or contribute to, a variety of pathological conditions, including:

(a) sustained coronary and cerebral vasoconstriction;
(b) hypertension;
(c) bronchospasm;
(d) atherosclerosis (endothelin is mitogenic [Chapter 15]);
(e) neurological dysfunction;
(f) renal failure, and
(g) myocardial infarction.

Factors Which Stimulate the Expression of Preproendothelin mRNA

One of the intriguing properties of endothelin-1 is that its production is enhanced under conditions which accompany endothelial cell injury. Shear stress, β-transforming growth factor and thrombin (Table 9.4) provide examples of conditions and agents which stimulate endothelin-1 production. Other stimulating agents include adrenaline, the Ca^{2+} ionophore A23187 and interleukin-1, a cytokine which is released by macrophages (Yoshizumi et al. 1990) (Table 9.4).

Table 9.4. Some factors which stimulate the expression of preproendothelin mRNA

Agent	Tissue	Reference
Thrombin (2 μ per ml)	Cultured porcine aortic endothelial cells	Yanagisawa et al. 1988a
Ca^{2+} ionophore (A23187) (10^{-6} M)	Cultured porcine aortic endothelial cells	Yanagisawa et al. 1988a
Adrenaline (10^{-6} M)	Cultured porcine aortic endothelial cells	Yanagisawa et al. 1988a
Transforming growth factor β_1	Cultured porcine aortic endothelial cells	Yanagisawa et al. 1988a
Physical shear stress	Cultured porcine aortic endothelial cells	Yoshizumi et al. 1989
Phorbol ester (0.5 M)	Cultured porcine aortic endothelial cells	Inoue et al. 1989
Myocardial infarction	Humans	Miyauchi et al. 1989
Interleukin-1	Cultured endothelial cells	Yoshizumi et al. 1990

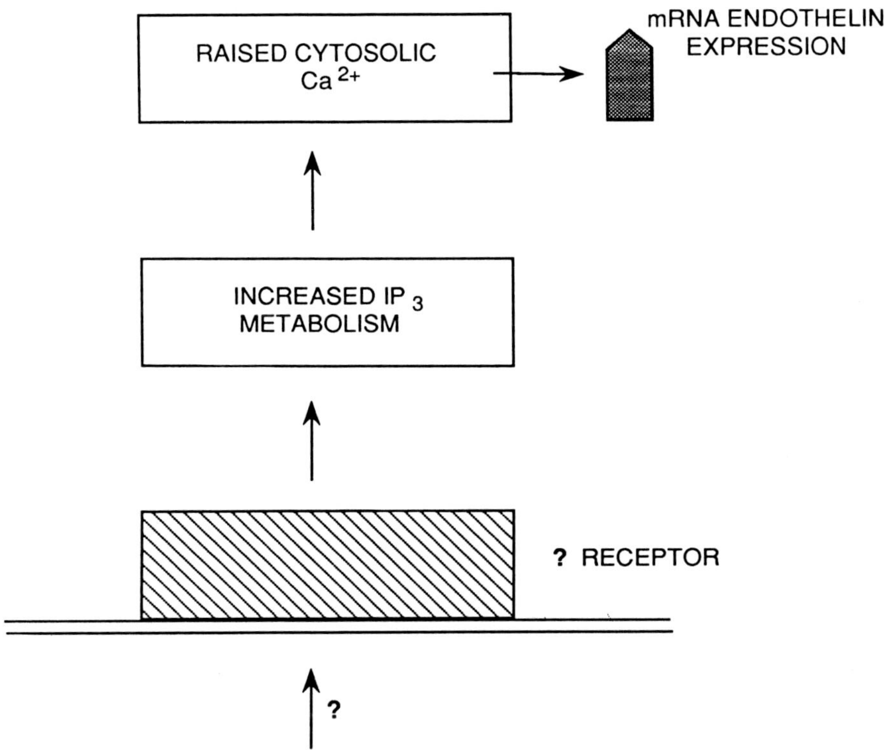

Fig. 9.1. Possible involvement of phosphoinositol (P1) metabolism in expression of endothelin mRNA

Evidence based on experiments in which the Ca^{2+} ionophore ionomycin was used to stimulate endothelin-1 mRNA expression has given rise to the idea that the expression of the mRNA for endothelin-1 may be controlled by a phosphoinositol-induced rise in intracellular Ca^{2+} (Figure 9.1). The response is relatively rapid. In the case of the Ca-ionophore-induced expression of the gene, up-regulation is evident within ten minutes (Inoue et al. 1989a, b). Phorbol-ester induced up-regulation occurs equally quickly.

Conditions Which Stimulate Endothelin Secretion

Three factors determine the circulating and tissue levels of endothelin-1. They are:
(a) the rate of expression of endothelin mRNA;
(b) the rate of endothelin secretion; and
(c) its rate of removal.

In marked contrast to the atrionatriuretic factor, storage granules do not seem to exist for the endothelins. Hence there must be a continuous cycling of production,

Table 9.5. Factors which stimulate endothelin secretion

Agent	Reference
Arginine-vasopression (AVP) (10^{-7}–10^{-6} M)	Emori et al. 1989a
Angiotensin II (10^{-8}–10^{-6} M)	Emori et al. 1989a
Ionomycin (Ca^{2+} ionophore) (10^{-6} M)	Emori et al. 1989a
Phorbol ester (10^{-6} M)	Emori et al. 1989a
Thrombin (0.2–20 U · ml^{-1}) (1.0 U · ml^{-1})	Schini et al. 1989
Water deprivation (4 days)	Yoshizawa et al. 1990
Endotoxin	Sugiura et al. 1989b
	Pernow et al. 1989a

The thrombin-induced stimulation of endothelin-1 secretion is both time (Schini et al. 1989) and concentration (Emori et al. 1989) dependent and is attenuated in a dose-dependent manner by cyclohexamide. The stimulatory effect of angiotensin II and arginine-vasopressin is blocked by the relevant receptor blockers

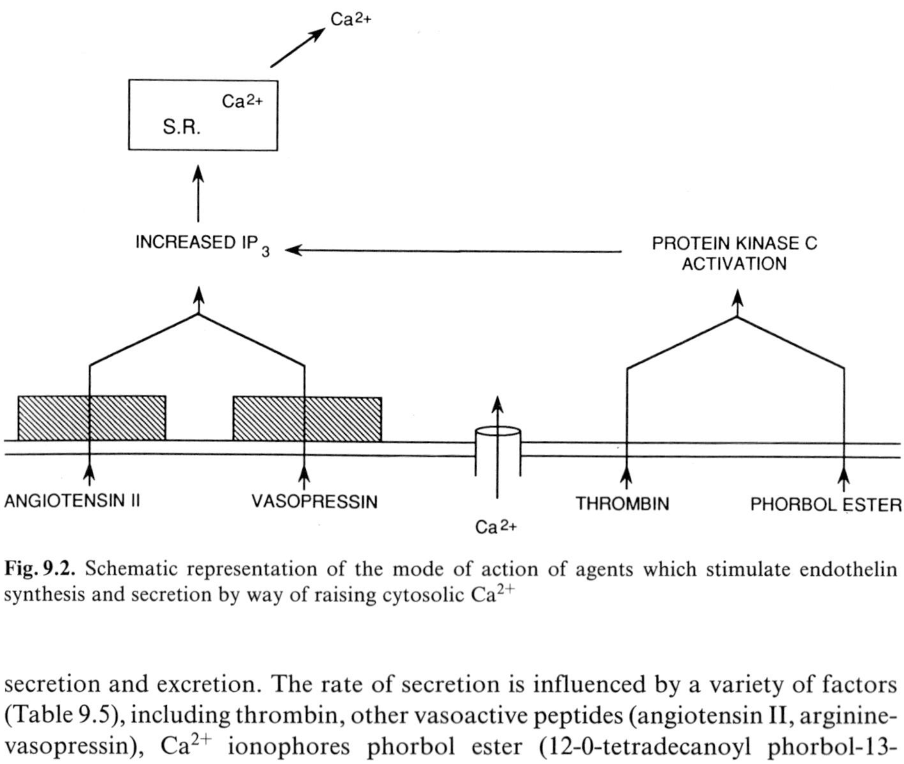

Fig. 9.2. Schematic representation of the mode of action of agents which stimulate endothelin synthesis and secretion by way of raising cytosolic Ca^{2+}

secretion and excretion. The rate of secretion is influenced by a variety of factors (Table 9.5), including thrombin, other vasoactive peptides (angiotensin II, arginine-vasopressin), Ca^{2+} ionophores phorbol ester (12-0-tetradecanoyl phorbol-13-acetate) and water deprivation. The thrombin-induced stimulation is both time (Schini et al. 1989) and concentration (Emori et al. 1989a) dependent and is attenuated in a dose-dependent manner by cyclohexamide. The stimulatory effect of angiotensin II and arginine-vasopressin is also dose-dependent and in both cases is blocked by the relevant receptor antagonists – [Sar^1,Ala^8]-angiotensin II for the angiotensin receptor, and [1-β-mercapto-(β,β-cyclopentamethylene proprionic acid), 2-0-methyltyrosine] (POMP)-AVP for the vasopressin receptor. On balance, therefore, it seems probable that the *secretion* of endothelin-1 is controlled by receptor (angiotensin II, vasopressin) and Ca^{2+}-mediated pathways (Figure 9.2), with Ca^{2+} providing the final translocating signal for mRNA expression.

Clearance of Endothelin from the Circulation

Although endothelin disappears from the circulation with a half-life of around eighty seconds it is stable for at least thirty minutes when added to isolated blood or plasma at 37°C. Specific removal systems must therefore exist. So far two possible routes have been identified – through the kidneys (Pernow et al. 1989b) and through the lungs (de Nucci et al. 1988), but as yet it is not known whether such removal involves either:

(i) metabolic degradation;
(ii) uptake into endothelial or other cell types; or
(iii) receptor binding and subsequent degradation.

Two findings point towards the kidney being a major route of clearance:

(i) the plasma endothelin levels increase dramatically in uraemic patients (1 pg per ml for controls compared with up to 35 pg per ml for uraemic patients (Chapter 10)); and
(ii) the urinary levels are higher (33.1 pg per ml) (Berbinschi and Ketelslegers 1989) than might otherwise have been expected.

Summary

1. Endothelin-1 is widely distributed throughout the body, with relatively high concentrations in the inner medulla of the kidney, in the lungs and in certain parts of the brain.
2. Under non-pathophysiological conditions the plasma levels of endothelin-1 are relatively low – so low, in fact, that it's significance under non pathological conditions is questionable. Relatively high levels of endothelin-1 are present in human milk. The level in milk > urine > plasma.
3. The synthesis of endothelin is not restricted to the endothelium: it also occurs in many other tissues, including the kidneys, the spinal cord and in certain areas of the brain.
4. The expression of mRNA encoded for endothelin and the secretion of endothelin is enhanced under conditions which accompany cell injury and emergency conditions. Thrombin, β growth transforming factor, adrenaline, vasopressin, angiotensin, increased shear stress and water deprivation all act in this way.
5. Removal of endothelin from the circulation occurs with a half life of a little over a minute. Kidneys and lungs are probable sites of removal, and receptor binding may be involved.

Chapter 10

Effect of Various Pathological Conditions on Plasma Endothelin

"The chess-board is the world; the pieces are the phenomena of the universe; the rules of the game are what we call the laws of Nature."
T. H. Huxley, in: Chapter III, Lay Sermons.

There can no longer be any doubt as to whether endothelin-1 is a normal component of plasma (Chapter 9), although its relatively late detection (Hartter and Woloszczok 1989); Miyauchi et al. 1989; Ando et al. 1989; Saito et al. 1989a; Cernacek and Stewart 1989) provides some idea of the technical difficulties which were encountered in developing a suitable, specific and sufficiently sensitive radioimmune assay. This difficulty was compounded by two factors:
(i) the plasma levels are normally low; and
(ii) plasma albumin interferes with the assay (Cernacek and Stewart 1989).

Compared with the plasma levels, the urinary levels of endothelin-1 are relatively high (< 1 and > 30 mg per ml respectively). Presumably the high urinary levels reflect the role played by the kidneys in removing the polypeptide from the circulation (Pernow et al. 1989b).

The presence of endothelin-1 in plasma raises some interesting questions (even though the circulating levels are low), because it is such a potent coronary and systemic vasoconstrictor (Kurihara et al. 1989a, b; Clarke et al. 1989; Ezra et al. 1989; Clozel and Clozel 1989). It is also arrhythmogenic and can cause sudden death (Terashita et al. 1989), as well as being mitogenic (Hirata et al. 1989d).

There are at least three reasons why endothelins might be present in the circulation.
(i) They may be circulating hormones involved in maintaining normal vascular tone. This applies particularly to endothelin-1. At any time, however, the direct vasoconstrictor effect of endothelin-1 may be partially, or even completely, offset by the release of vasodilator agents – including (Figure 10.1) the atrionatriuretic peptide, ANP, and the endothelial derived prostanoids, as well as EDRF. This possibility poses another interesting question as to what happens if this compensatory release of vasodilator substances fails, and the constrictor effect of the endothelin remains unopposed? Alternatively,
(ii) the presence of endothelin-1 in the plasma may simply reflect a leakage or overflow of the polypeptide from areas where it is being produced to act locally (Figure 10.2). If this is the case then it is the interstitial rather than the plasma levels which should be being measured.

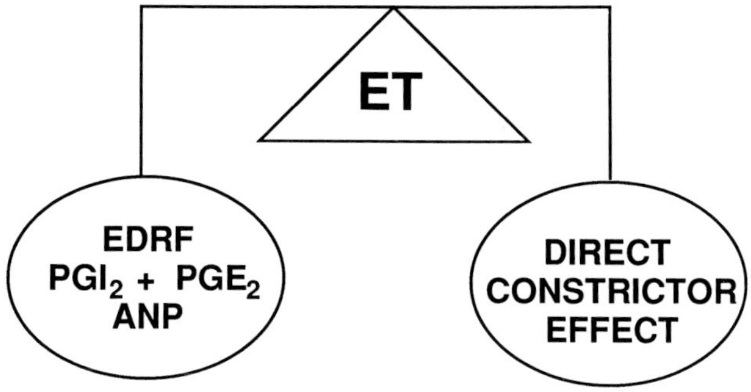

Fig. 10.1. Schematic representation of the balancing effect which released vasodilator substances must exert on the direct constrictor effect of endothelin-1. EDRF is endothelial-derived relaxing factor. PGI_2 and PGE_2 are locally released prostanoids. ANP is the atrionatriuretic peptide (or factor) released from its storage granules in the atria in response to stretch and other stimuli. EDRF, PGI_2 and PGE_2 are released from the endothelium

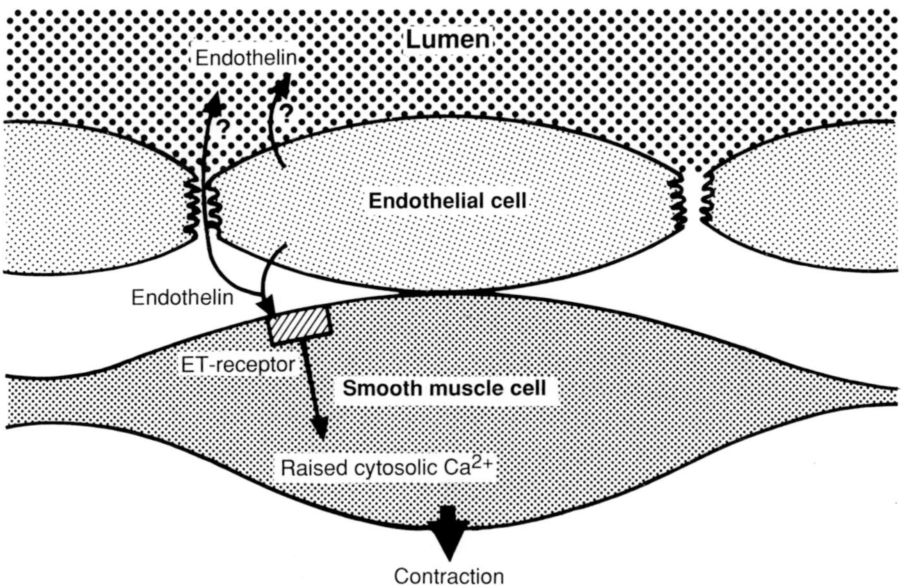

Fig. 10.2. Schematic representation of the "spill over" of endothelin into the circulation. The target site for the released endothelin is the endothelin (ET) receptor on the smooth muscle cell. Activation of this receptor results in a sustained rise in cytosolic Ca^{2+}, and hence constriction

(iii) The polypeptide may only be produced in significant amounts in response to a local emergency, including vascular injury.

Irrespective of which if any of these conditions apply both the rate of endothelin-1 production and its rate of clearance from the circulation, need to be controlled if a

Table 10.1. Conditions which modify plasma endothelin-1

Condition	Endothelin-1 (pg per ml)		
	Normal level	Patient level	Reference
A. Patient Studies			
Surgery	1.5	7.3	Saito et al. 1989b
Surgery	1.5	150–500% increase	Hirata et al. 1989b
Subarachnoid haemorrhage	1.5	12.0	Masaoka et al. 1989
Cardiogenic shock	0.26	3.65	Cernacek + Stewart 1989
Pulmonary hypertension	0.26	1.52	Cernacek + Stewart 1989
Renal dialysis	0.26	1.52	Cernacek + Stewart 1989
Renal failure	1.5	10.4	Tomita et al. 1989
Uraemia	< 3.5	12.3	Koyama et al. 1989
Renal transplantation	5	15	Berbinschi and Ketelslegers1989
Myocardial infarction	2.9	8	Salminen et al. 1989
Myocardial infarction	1.5	3.2	Miyauchi et al. 1989
B. Animal Studies			
Endotoxic shock (in swine)	23	160	Pernow et al. 1989a
Endotoxic shock (in sheep)	19.7	56.1	Morel et al. 1989

lethal event is to be avoided (Ezra et al. 1989; Terashita et al. 1989). Perhaps this is the place to re-emphasise how potent endothelin-1 is as a vasoconstrictor agent. In humans, for example, it's infusion into the forearm at a rate of 11 pg per minute causes a thirty percent reduction in blood flow (Webb and Cockcroft 1989). Readers might be tempted to think that because the polypeptide has a relatively short half life (77 seconds, Pernow et al. 1989b) its presence in the circulation does not provide cause for concern. This would be an erroneous conclusion, however, because the prolonged binding of the polypeptide to its receptor (Chapter 7) ensures that it evokes a sustained constrictor response which persists long after the polypeptide will have been cleared from the circulation (Clarke et al. 1989). Admittedly negative feed-back systems do exist – including the "down" regulation of the angiotensin system, the "down" regulation of the endothelin receptor (Chapter 7) and the local and general release of various vasodilator agents (Figure 10.1) – but the overall response is that of a dose-dependent intense and sustained vasoconstriction.

As yet little information is available concerning the mechanisms which regulate the production and accumulation of endothelin-1 in the plasma. Nor is there any irrefutable evidence to show that altered plasma levels are causally linked to specific pathological conditions. However abnormally high plasma levels do occur under pathological conditions which are normally associated with a raised vascular resistance in one or more of the vascular beds. These conditions include surgical stress, subarachnoid haemorrhage, cardiogenic shock, acute myocardial infarction, pulmonary and systemic hypertension, uraemia, renal failure, renal dialysis, cold stress and transplantation (Table 10.1).

Plasma endothelin-1 levels are not the only levels which change – altered levels in lymph, in cerebrospinal fluid and even in bronchial exudates have also been observed.

The purpose of this chapter is simply to provide some idea of the magnitude of these various changes and of the conditions under which they occur.

Plasma Levels of Immunoreactive Endothelin-1 in Various Pathological Conditions

Bearing in mind the advice given to Alice (Chapter 1) regarding the desirability of starting at the beginning and continuing until the end ... it might be useful to recall that plasma levels of immunoreactive endothelin-1 in healthy human volunteers range between 0.3 and 1.7 pg per ml (Table 9.2). The levels remain constant throughout the day, are not sex-related (Saito et al. 1989a) and do not increase in response to, or in association with, a rise in blood pressure caused by physical exercise (Hartter et al. 1989). The levels are affected by postural changes and body fluid status, however. For example the plasma levels are higher in the upright relative to the supine position (Shichiri et al. 1990), and infusions of saline cause the levels to fall. These changes closely mimic the changes in plasma renin which occur under the same conditions.

Surgery

Surgical interventions (abdominal surgery and resection of tumours) produce an increase in plasma endothelin-1. In two separate studies (Saito et al. 1989b; Hirata et al. 1989b) increased levels were detected (Table 10.1) immediately after completion of surgery but they returned to control levels within two or three days.

This surgically-induced increase is probably due to:

(i) augmented secretion and leakage of the polypeptide from vascular endothelial cells injured during the surgical manipulations; and
(ii) "stress"-provoked increase in production (Yanagisawa et al. 1988a).

Although the plasma levels of around 7 pg per ml which occur under these conditions is below the level needed to provoke vasoconstriction (Chapter 5, and Yanagisawa et al. 1988a) levels in the immediate vicinity of the traumatized vessels may be much higher and therefore closer to the threshold concentration needed to provoke a constrictor response.

Aneurysmal Subarachnoid Haemorrhage

The onset of subarachnoid haemorrhage is accompanied by substantial increases in plasma endothelin-1, and, in contrast to the relatively short-lived increase caused by surgical interventions, the increase persists for days, particularly in patients with symptomatic evidence of cerebral vasospasm (Masaoka et al. 1989). In these patients plasma levels of 12 pg per ml are not uncommon and may even be responsible for the long lasting vasospasm which so often occurs under these conditions.

Cardiogenic Shock

In patients with cardiogenic shock (Cernacek and Stewart 1989) the plasma levels of endothelin-1 increase by more than a factor of ten. This increase contrasts with the *absence* of any change in patients with stable cardiac failure (Cernacek and Stewart 1989), and in experimentally-induced acute cardiac failure induced by rapid right ventricular pacing (Cavero et al. 1989).

Hypertension

Plasma endothelin-1-like immunoreactivity increases during pulmonary (Cernacek and Stewart 1989) and essential hypertension (Saito et al. 1990). However, in the patients with essential hypertension, where the plasma levels often double, the increase does not correlate with either the systolic or diastolic blood pressure, or heart rate. Probably, therefore, the increase is due to vascular complications of the disease, rather than being causally related to its onset.

Myocardial Infarction

Here the plasma endothelin-1 levels increase during the first hours of the ischaemic episode (Figure 10.3) (Salminen et al. 1989; Miyauchi et al. 1989). Moreover the increase is accompanied by increased plasma levels of "big" endothelin-1 (Miyauchi et al. 1989), indicating an increased rate of synthesis. It follows, therefore that during acute episodes of myocardial ischaemia:
(i) plasma endothelin-1 levels rise (by as much as two hundred and twenty percent during the first day); and
(ii) this increase is accompanied by a parallel increase in endothelin synthesis (Miyauchi et al. 1989).

The cause of these increases is unknown. Possibly platelet aggregation may be involved, because this could trigger the local accumulation of thrombin and transforming growth factor -β (TGF-β), both of which induce endothelin-1 gene expression (Chapter 9, and Yanagisawa et al. 1988a).

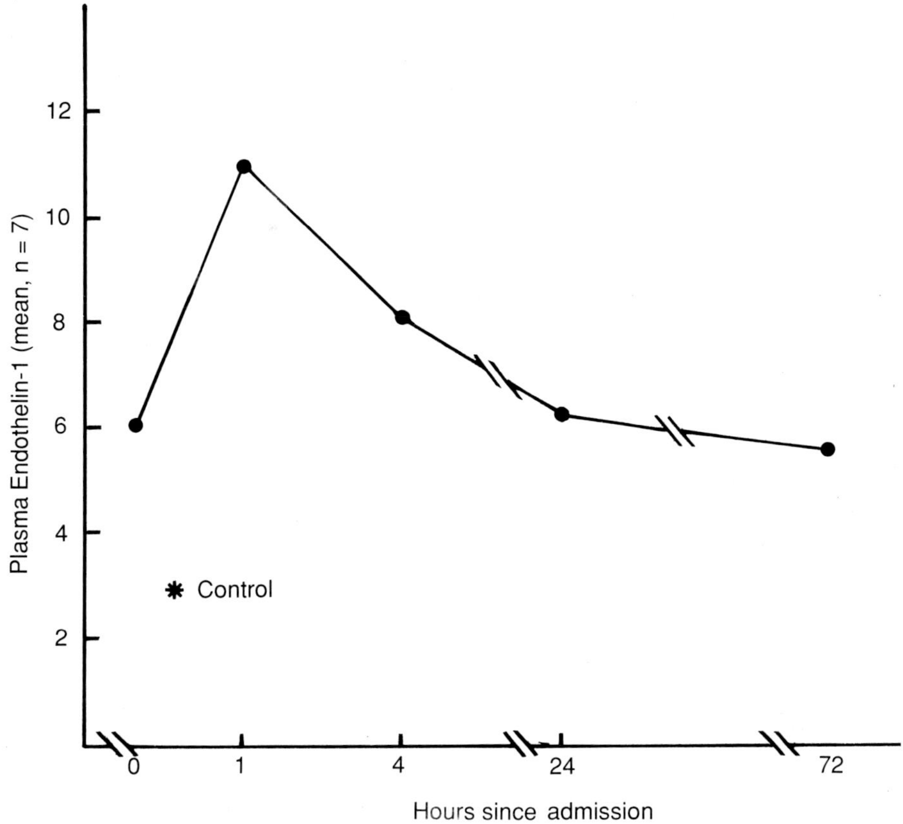

Fig. 10.3. Changes in plasma endothelin-1 levels in the early stages of myocardial infarction. Data taken from Salminen et al. 1989

Uraemia, Acute Renal Failure and Dialysis

Uraemia, acute renal failure and dialysis are all accompanied by significant rises in plasma endothelin-1 (Table 10.1). It might be tempting to dismiss these rises as a secondary consequence of reduced clearance through the kidneys but such an idea is not plausible, because in *chronic* renal failure the plasma levels are unchanged (Tomita et al. 1989).

Kidney Transplantation

This is another condition under which the level of endothelin-1 in the plasma rises, with a mean level of 15 pg per ml persisting for as long as forty eight hours after transplantation (Berbinschi and Ketelslegers 1989).

Experimental Conditions Which Provoke Raised Plasma Levels of Endothelin-1

Although artificially induced endotoxic shock caused by the intravenous administration of E coli does not strictly qualify for inclusion in a chapter which is devoted to the consideration of pathological conditions which are associated with altered plasma levels of endothelin-1 (Table 10.1) the results obtained from such experiments are of interest, particularly since the accompanying physiological response includes a sustained increase in vascular resistance. In both sheep (Morel et al. 1989) and swine (Pernow et al. 1989a) the plasma endothelin levels increase substantially (from around 4 pg per ml to 29 pg per ml in the swine, and from around 3 pg per ml to around 10 pg per ml in the sheep). The most likely cause of the rise is endotoxin-induced cell injury. The time course of the response (Pernow et al. 1989a) parallels the increase in splenic and renal vascular resistance.

Cold Provocation

Simply immersing one hand into ice-cold water for two minutes is sufficient to cause an immediate rise in the plasma endothelin-1 levels in the immersed hand, followed, some minutes later, by raised levels in the contralateral hand (Fyhrquist et al. 1989).

Other Changes

Cerebrospinal Fluid

The presence of endothelin-1 in the spinal cord and brain (Chapter 9) has given rise to the idea that it might function as a neuropeptide. Cerebrospinal fluid also contains endothelin-1, usually at a concentration of around 8 pg per ml (Hoffman et al. 1989), which is about eight times higher than normal plasma levels. In patients with major depression the levels are substantially lower (3.8 pg per ml, a fall of approximately fifty percent).

Bronchial Exudates

The presence of endothelin-1 in bronchial exudates is predictable, since tracheal epithelial cells secrete endothelin-1 (Black et al. 1989). Moreover an increased level in asthmatics might have been expected, since tissue injury and inflammation stimulate endothelin mRNA expression (Yanagisawa et al. 1988a). The levels do rise during status asthmaticus (Nomura et al. 1989), from a control level of 0.05 pg per ml of lavage fluid to around 0.3 pg per ml during the attack – *an increase of five hundred percent.* This would seem to be a "dog chasing its tail" effect, because endothelin-1 is a potent bronchoconstrictor (Uchida et al. 1988) even in endothelial-denuded preparations (Hay et al. 1989).

Summary

1. In normal healthy volunteers the levels of endothelin-1 in the plasma, cerebrospinal fluid and bronchial exudates are low, and hence below the threshold needed to evoke a constrictor response.
2. Under certain pathological conditions, including myocardial infarction, acute renal failure, haemodialysis, cardiogenic shock, hypertension, cold stress and subarachnoid haemorrhage the plasma endothelin-1 levels rise during the acute stages of the episode. Major surgical interventions produce a similar response.
3. In contrast to acute myocardial ischaemia, neither stable cardiac failure, nor experimentally-induced acute cardiac failure evokes an increase in plasma endothelin-1.
4. Levels of endothelin-1 in the cerebrospinal fluid are abnormally low in patients with depressive mental illness. Conversely, the levels in bronchial exudates increase during status asthmaticus.
5. The infusion of E coli causes a substantial and time-dependent increase in plasma endothelin-1, accompanied by sustained vasoconstriction.
6. The conditions under which plasma and bronchial exudate endothelin-1 levels increase all share one common feature – that of cell injury, a factor which prompts the question as to whether or not tissue injury is the primary signal for increased endothelin production and "spill over" into the plasma, cerebrospinal fluid and bronchial exudates.

Chapter 11

Endothelin and the Kidney

"We used to think that if we knew one, we knew two, because one and one are two. We are finding that we must learn a great deal more about "and".
SIR. A. EDDINGTON

As soon as the structure and basic properties of endothelin-1 had been determined investigators began studying its effect on specific organs. The kidneys, the brain and the heart were natural targets for attention. As far as the kidneys are concerned it soon became apparent:

(i) that renal arteries are particularly sensitive to the polypeptide – more sensitive even than the systemic vasculature (Pernow et al. 1989c);
(ii) that the kidney contains specific high affinity binding sites for the polypeptide (Jones et al. 1989a, b; Kohzuki et al. 1989);
(iii) that the distribution of these binding sites throughout the kidney is non-homogeneous (Kohzuki et al. 1989; Davenport et al. 1989b);
(iv) that the kidney contains mRNA for endothelin-1 and accordingly is capable of synthesizing the polypeptide (MacCumber et al. 1989), and
(v) that when injected into the renal artery, endothelin-1:
    (a) reduces renal blood flow,
    (b) increases renal artery resistance,
    (c) slows glomerular filtration,
    (d) reduces urinary sodium excretion, and
    (e) stimulates prostaglandin E_2 and I_2 release (Miura et al. 1989). In addition,
    (f) whereas high doses of endothelin-1 increase renal arterial resistance, low doses can cause dilation (Lippton et al. 1988; Figure 11.1). Under in vivo conditions, therefore, the response is often biphasic, with a transient elevation of renal blood flow followed by a gradual but sustained fall (Nakamoto et al. 1989). Finally,
(vi) the constrictor effect of endothelin-1 on the renal vasculature, and particularly the afferent arterioles, is markedly attenuated by the dihydropyridine-based calcium antagonists (Loutzenhiser et al. 1990).

The Hypersensitivity of Renal Arteries to Endothelin-1

There are at least three reasons for arguing that the renal vasculature is particularly sensitive to the vasoconstrictor effect of endothelin-1.

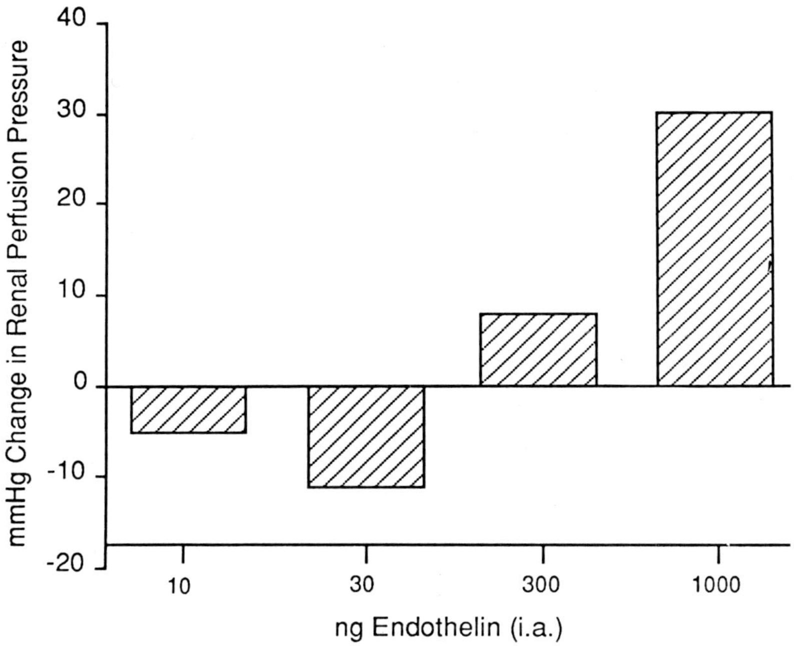

Fig. 11.1. Effect of injecting 10, 30, 300 and 1000 ng endothelin-1 into the renal artery of anaesthetized cats. Results are from Lippton et al. 1988, and are the mean of seven experiments at each dose level. The endothelin-1 was given as a bolus

Firstly, concentrations of endothelin-1 which are below the threshold needed to cause a rise in mean arterial blood pressure reduce renal blood flow. Gardiner et al. described such a response in unrestrained, chronically instrumented normotensive rats (Gardiner et al. 1989a). Others have used awake rats (Madeddu and Glorioso 1989) or anaesthetized pigs (Pernow et al. 1989c) and rats (Hirata et al. 1989c). In each case the response has been the same, with

(i) endothelin constricting the renal vasculature at a dose-level which is *below* the threshold for the systemic vasculature;
(ii) the effect develops slowly (relative to that of angiotensin II) but is sustained; and
(iii) is often biphasic, with the constrictor response being preceded by a transitory dilation.

Secondly, when injected into rabbits the endothelin antibody decreases renal vascular resistance, without changing mean arterial pressure (Miyamori et al. 1989). Such a response would be anticipated only if the renal vasculature is hypersensitive.

Thirdly, studies in which either isolated perfused kidneys (Firth et al. 1988) or isolated renal arteries (Tomobe et al. 1988) have been used have allowed direct comparisons to be made between the renal vasoconstrictor potency of endothelin-1 and that of other constrictor agents, including angiotensin II. These studies have shown:

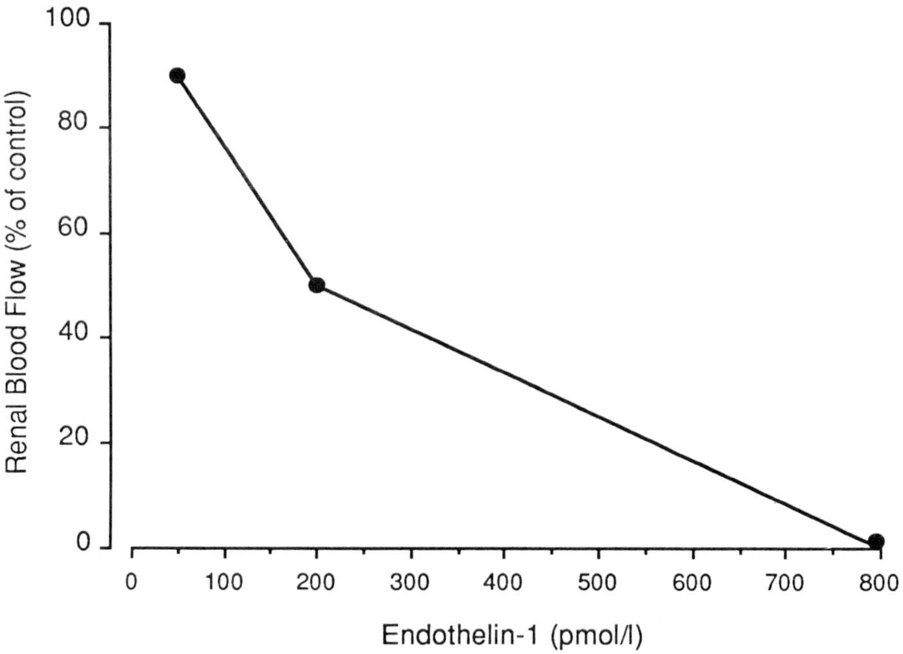

Fig. 11.2. Effect of endothelin-1 on renal blood flow in isolated perfused rat kidney. (Data from Firth et al. 1988)

(i) that the renal constrictor effect of endothelin-1 is dose-dependent (Figure 11.2); and
(ii) that whereas 200 pmol/l of endothelin-1 is sufficient to cause a fifty percent reduction in renal blood flow, five times as much angiotensin II (1000 pmol/l) is needed to produce an equivalent response – even though prior to the discovery of the endothelins angiotensin II was the most potent known renal vasoconstrictor (Firth et al. 1988).

These studies have provided other information concerning the renal constrictor effect of endothelin-1. For example, the time course of the constrictor response is quite different from that of angiotensin II. In the latter case the response peaks before ten minutes and recovery is complete within sixty minutes. In the case of endothelin-1 the maximum response does not occur until thirty minutes after the bolus injection has been given, and the response persists well beyond ninety minutes.

Although the renal constrictor effect of endothelin-1 is sustained, at least two interventions can be introduced to relax the endothelin-constricted arteries. These are:
(i) the addition of isoprenaline (Tomobe et al. 1988), and
(ii) the infusion of atrial natriuretic peptide (Hirata et al. 1989c).

The fact that the atrial natriuretic peptide (ANP) relaxes endothelin-1 constricted renal arteries is of considerable importance, because endothelin-1 stimulates ANP release. This, coupled with the fact that endothelin-1 also stimulates the release of certain autacoids, including PGI_2 and PGE_2 (but not thromboxane A_2) (Miura et al. 1989; Rae et al. 1989) from the kidney, and given that the kidney generates substantial amounts of these vasodilator autacoids in response to constrictor stimuli, *means that at least two feedback mechanisms exist to control the constrictor effect of endothelin-1 on renal blood vessels*. To this must be added at least one other vasodilatory mechanism – the release of EDRF (de Nucci et al. 1988).

In summary therefore, although endothelin-1 is a potent constrictor of the renal vasculature, under *in vivo* conditions the constrictor response is ameliorated by

(i) the release of ANP from the cardiac atria;
(ii) the release of vasodilatory autacoids from the kidney, and
(iii) the release of EDRF from the endothelium of the vasculature.

The existence of such a feed-back mechanism raises at least two questions: does the condition ever arise under which the amount of endothelin-1 which is available is sufficient to override these feed-back dilator mechanisms, and *secondly*, does the failure of these control mechanisms contribute to the aetiology of pathological conditions which are associated with renal artery stenosis? The answer to the first of these questions is probably in the affirmative, since injections of the endothelin-antibody increases renal blood flow in normotensive rabbits, without altering mean arterial blood pressure (Miyamori et al. 1989). Such a response is only possible if renal vasculative tone was already being determined, in part at least, by the presence of endothelin-1. This need not necessarily be the plasma endothelin-1, since kidneys are capable of synthesizing the polypeptide in situ (MacCumber et al. 1989). Whether these feed-back mechanisms fail under pathological conditions is currently under intensive investigation.

The Heterogeneity of the Renal Vascular Effects of Endothelin-1

The constrictor effect of endothelin-1 on the renal vasculature must be a *direct* effect, since it persists in ganglion-blocked preparations (Wright and Fozard 1988). Moreover, the response is heterogeneous throughout the vascular field, being *more prominent in the afferent compared with the efferent arterioles* (Loutzenhister et al. 1990). It is also prominent in the arcuate and interlobular arteries (Hirata et al. 1989c). The renal vasculature usually exhibits a heterogeneous response to vasoactive substances. Angiotensin, for example, primarily alters efferent arteriolar blood flow in the renal circulation. By contrast, dopamine preferentially dilates the preglomerular vessels.

Precisely why endothelin-1 predominately affects the preglomerular renal vessels is unknown but it may be linked to the fact that the tone of these vessels is highly dependent on extracellular Ca^{2+} (Cohen and Fray 1982). This may also explain why the renal vascular is hypersensitive to endothelin-1, relative to the systemic vasculature.

In general therefore, endothelin-1 is a particularly potent constrictor of the renal vasculature, particularly the preglomerular vessels. This constriction is accompanied by a reduction in renal blood flow, a reduced glomerular filtration rate and a slowed urinary excretion. Since endothelin promotes ANP release, and since ANP dilates the preglomerular vessels (Hirata et al. 1989c) and has a natriuretic effect, it can be anticipated that ANP will attenuate some of the renal effects of endothelin-1, including the endothelin-1 induced contraction of the preglomerular blood vessels and its inhibitory effect on Na^+ excretion. Such a feed-back mechanism should come into play *before* mean arterial pressure rises, since endothelin-1 has a *direct* stimulant effect on the secretion of ANP from cardiac atrial cells (Fukuda et al. 1988).

The Distribution and Properties of Endothelin-1 Receptors in the Kidney

As mentioned in Chapter 7, the kidneys are well endowed with endothelin-1 binding sites. Their distribution, however, is heterogeneous (Figure 11.3). The glomeruli, inner medulla and the vasa rector bundles are areas of high density binding (Davenport et al. 1989b; Kohzuki et al. 1989; Hoyer et al. 1989), but the outer medulla is relatively deficient (Table 11.1). The characteristics of these binding sites

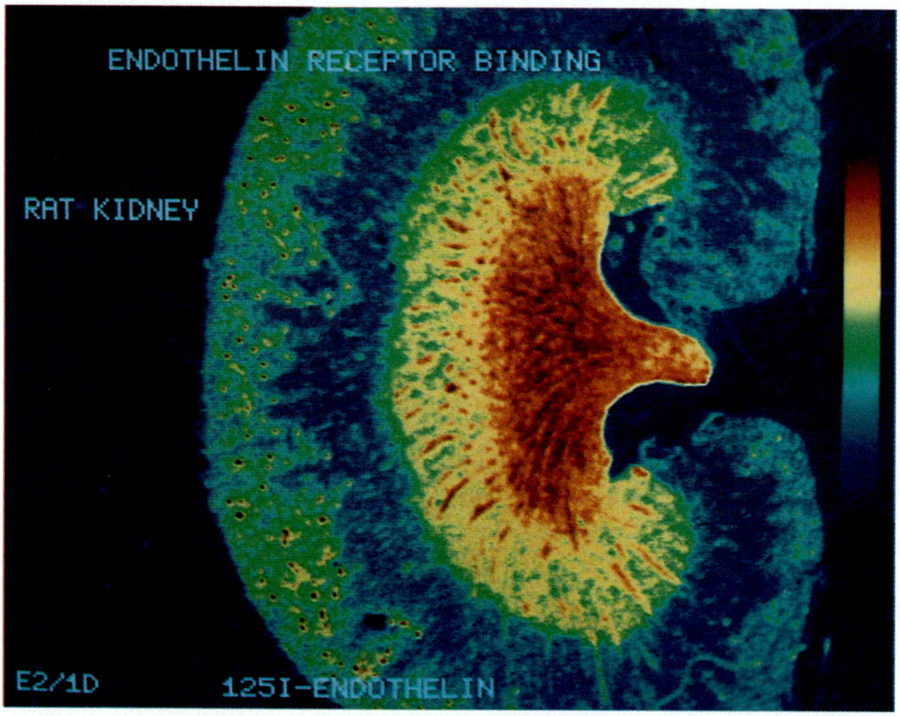

Fig. 11.3. Autoradiograph of [^{125}I]endothelin-1 binding sites in the rat kidney

Table 11.1. Distribution of [^{125}I]endothelin-1 binding sites in rat and monkey kidneys

Tissue	[^{125}I]-endothelin-1 binding
Rat (1)	*fmol/mm²*
Glomeruli	377 ± 75
Inner medulla	322 ± 68
Vasa recta bundles	339 ± 58
Outer medulla	N.D.
Monkey (2)	*fmol/mg tissue*
Inner medulla	364
Outer medulla	16

1. Data from Kohzuki et al. 1989
2. Data from Hoyer et al. 1989

The rank order of binding is glomeruli > inner medulla = vasa recta bundles >>> outer medulla
N.D. = not detectable

closely resemble those described in Chapter 7 for other tissues. The binding is saturable and the sites are *insensitive* to other vasoactive peptides (angiotensin I and II, ANP and vasopressin), to calcium antagonists (verapamil, nifedipine and diltiazem) (Kohzuki et al. 1989) and to inhibitors of the Na^+ channels (α scorpion toxin) and the N-type Ca^{2+} channels (w conotoxin GVIA).

There are several consequences of the non-homogeneous distribution of these endothelin binding sites:

(i) the high density of binding in the glomeruli is in keeping with the idea that endothelin-1 modulates glomerular function;
(ii) the high density of binding in the vasa recta bundles agrees well with the idea that endothelin-1 regulates renal medullary blood flow;
(iii) although the density of binding in the inner medulla is high, the cellular location of these sites is not known, as yet. Possible cellular sites include the loop of Henle and the collecting ducts.
(iv) The distribution of the endothelin-1 binding sites in the kidney closely resembles that of the angiotensin II and ANP receptors (Table 11.2). For instance, for both endothelin-1 and ANP receptors the density of binding is high in the glomeruli, the vasa recta bundles and the inner medulla (Table 11.2). Angiotensin II receptor binding sites are also localized in the glomeruli, vasa recta bundles and in the proximal tubules (Chai et al. 1986; Mendelsohn et al. 1987). The similar pattern of distribution of these binding sites may be coincidental. Alternatively, it may mean that the regulation of renal haemodynamics, glomerular function and proximal tubular reabsorption involves all three vasoactive peptides (ANP, angiotensin II and endothelin-1) acting in concert.

In the rat kidney binding occurs with a K_D (concentration required to occupy half of the binding sites) of 0.48 ± 11 nM for the glomeruli, 0.68 ± 0.19 nM for the inner

Table 11.2. Relative distribution of endothelin-1, angiotensin II and atrionatriuretic peptide receptors in kidneys

Polypeptide	Endothelin-1 (ET-1)	Atrionatri-uretic peptide (ANP)	Angiotensin II (Ang II)
Binding Site			
Glomeruli	High	High	High
Vasa recta bundles	High	High	High
Inner medulla	High	High	
Proximal tubules	Moderate	Moderate	Moderate

Data from Kohzuki et al. 1989

medulla, and 0.55 ± 0.11 nM for the vasa recta bundles. These values are approximately the same as those obtained for the heart, and are in quite reasonable agreement with the concentrations of endothelin-1 which reduce renal blood flow, decrease glomerular filtration rate, slow urinary sodium excretion and (Orita et al. 1989) stimulate PGE_2 production. Probably, therefore, these receptors are functional.

Endothelin-1 Binding Sites in Renal Glomerular Mesangial Cells

Mesangial cells are contractile, smooth muscle-like pericytes which, by contacting and dilating, help regulate glomerular haemodynamics. They are located in the central matrix of the glomerulus between capillary loops (Mené et al. 1989). As well as influencing glomerular filtration rate the mesangial cells also synthesize and assemble the mesangial matrix – an important function because this matrix helps to regulate the viscoelastic properties of the glomerular capillaries (Simonson and Dunn 1990).

Radioligand binding studies have shown that the mesangial cells contain two types of endothelin binding sites, with molecular weights of approximately 58,000 and 34,000 (Sugiura et al. 1989c). These binding sites display all the usual characteristics of the endothelin receptors (insensitivity to N- and L-type Ca^{2+} channel blockers, and insensitivity to other vasoactive polypeptides), and they can be divided into two groups on the basis of their affinity for [^{125}I]endothelin-1 – high affinity sites (ET_1) and low affinity sites (ET_2) (Badr et al. 1989). The role of the higher affinity sites is uncertain, but stimulation of the lower affinity sites leads to mesangial contraction secondary to a raised cytosolic Ca^{2+}, intracellular alkalinization, mitogenesis and c-fos protooncogene expression (Simonson and Dunn 1990; Badr et al. 1989). Since glomerular endothelial cells express mRNA for preproendothelin-1 and release an endothelin-like peptide (Marsden et al. 1989b), and the glomerular cells are contiguous with mesangial cells, it is highly probable that endothelin which is produced in and released from the glomerular cells activates the endothelin receptors on the surface of the contiguous mesangial cells, thereby initiating a train of events which results in mesangial contraction and proliferation. The former response allows mesangial cells to regulate, in part, glomerular filtration; the latter response may lead to cell remodelling, thereby *indirectly* regulating glomerular function.

Effect of Endothelin-1 on Renal Tubular Epithelial Function

Although endothelin is a potent constrictor of renal arteries, and although it usually slows urinary sodium excretion, it sometimes stimulates salt excretion (Zeidel et al. 1989). Presumably this is because the polypeptide inhibits the Na^+K^+ ATPase enzyme located in the inner medullary collecting duct cells (Zeidel et al. 1989). This inhibition seems to be specific for these cells, cells from the proximal tubules being unaffected. This inhibitory effect occurs at relatively low concentrations (below 0.1 nM), and these concentrations also stimulate PGE_2 synthesis. Since PGE_2 has a direct inhibitory effect on the Na^+K^+ ATPase enzyme it is possible that it mediates the inhibitory effect of the polypeptide. The tubular epithelial cells do contain functioning specific binding sites for endothelin (Neuser et al. 1990). Taken together these findings provide strong support for the hypothesis that endothelin contributes to the processes responsible for regulating epithelial reabsorption – quite apart from its direct effect on the renal vasculature and the secondary consequences of the resultant decrease in renal blood flow.

The in situ Production of Endothelin in the Kidney: Is It a Local Hormone?

mRNA for preproendothelin-1 occurs in the renal medulla but not in the glomeruli or the renal cortex (MacCumber et al. 1989). In the medulla, therefore, endothelin may be acting as a local hormone, since the medulla contains binding sites for the peptide (Table 11.2). The glomeruli also contain endothelin-specific binding sites, but since cells in this region of the kidney do not express mRNA for preproendothelin, they must rely primarily on circulating endothelin.

Endothelin-1 and Renal Failure: Is There a Link?

Given that endothelin-1 has such a profound and complex effect on the kidney it is quite likely that endothelin will be involved in some aspects of renal failure. Plasma levels are known to rise in uraemic patients but whether this is a primary or secondary effect is unknown. However, since the kidney is capable of expressing mRNA encoded for preproendothelin, measurements of tissue as well as plasma levels of endothelin will be needed before much progress can be made in answering this question. It is of interest to note, however, that the inner medullary levels of endothelin are abnormally low in spontaneously hypertensive rats (Kitamura et al. 1989b), possibly because of "down" regulation of mRNA endothelin expression.

Once again, however, it is impossible to interpret the clinical significance of this finding, because the reduction in medullary endothelin-1 concentration could be a compensatory mechanism.

Summary

1. The kidney contains endothelin-1 binding sites and their distribution coincides with that of angiotensin II and ANP receptors.
2. The receptors are not limited to the vasculature. High affinity binding sites are also present in the renal mesangial cells and the epithelial cells of the inner medullary collecting ducts.
3. mRNA for preproendothelin is present in renal glomerular cells and in the inner medulla. The polypeptide, therefore, may act on the kidney as a local hormone.
4. Endothelin causes a dose-dependent reduction in renal blood flow (preceded by a transitory dilation), a reduction in glomerular filtration rate, a reduction in urinary sodium excretion, and a stimulation of the renal secretion of PGI_2 and PGE_2.
5. Although the renal vasculature is highly sensitive to the constrictor effect of endothelin-1, negative feed back mechanisms involving the release of ANP, EDRF and PGI_2 and PGE_2 provide a counterbalance.
6. At relatively low concentrations (< 0.1 nM) endothelin-1 has an inhibitory effect on the Na^+K^+ ATPase activity in the inner medullary collecting duct cells.
7. The dihydropyridine-based calcium antagonists attenuate the constrictor effect of endothelin-1 on the renal vasculature.

Chapter 12

Endothelin and the Brain

"One sometimes finds what one is not looking for."
SIR ALEXANDER FLEMING

Endothelin-1 does not cross the blood-brain barrier (Koseki et al. 1989a) nor does it alter it's permeability (Feraci et al. 1989). It might have been assumed, therefore, that the endothelins would not influence the functioning of the central nervous system – apart from any effects which might arise either as a secondary consequence of a reduction in blood flow due to cerebral vasoconstriction, and vasospasm (Kauser et al. 1990), or effects which might occur because of an action on the circumventricular organs which have a deficient blood brain barrier. Once again, however, the endothelins are full of surprises, because not only do they act centrally – they are also produced locally.

Evidence of a Direct Central Effect of Endothelin (Endothelin-1)

There are several reasons for arguing that the endothelins have a direct central effect. Thus,

(i) when injected intracerebroventricularly endothelin-1 evokes a dose-dependent increase in mean arterial blood pressure, accompanied by an increase in heart rate and raised plasma levels of noradrenaline and adrenaline (Ouchi et al. 1989);
(ii) when injected into the right lateral ventricle endothelin-1 causes behaviourial disturbances which, although relatively slow in onset, involve ataxia and oscillatory side to side movements (Moser and Pelton 1989);
(iii) when injected intrastriatally endothelin-1 induces ischaemia-like lesions, particularly in the neostriatum (Fuxe et al. 1989);
(iv) when applied to rat spinal cord preparations endothelin-1 causes depolarization of the ventral root, indicating a direct effect on neuronal excitability (Yoshizawa et al. 1989b).

However, since the endothelins neither cross nor permeabilize the blood brain barrier changes caused by the central injection of or locally applied endothelin-1 will be nothing more than laboratory novelties unless the brain is actually capable of endothelin synthesis.

Table 12.1. Regional distribution of endothelin-1 (ET-1) and endothelin-3 (ET-3) in rat brain and other tissues

Region	fmol per gram tissue		
	ET-1	ET-3	ET-1/ET-3
Cerebral cortex	156.6	38.0	3.9
Cerebellum	100.4	60.6	1.7
Medulla oblongata	88.3	15.2	5.8
Pituitary gland	21.3	71.9	0.29
Lung	1,365.5	94.6	14.4
Heart	136.5	1.4	97.5
Intestine	309.2	174.2	1.77
Kidney	112.4	10.2	11.0

Note the relatively high proportion of ET-3, relative to ET-1 in the pituitary gland, and the relatively high concentration of ET-1 (relative to ET-3) in the heart. (Data from Matsumoto et al. 1989)

Evidence of the Local Production of Endothelins in the Brain

As can be seen from Table 12.1, the brain contains at least two isoforms of endothelin – endothelin-1 (Hoyer et al. 1989; Shinmi et al. 1989b; Yoshimi et al. 1989; Yoshizawa et al. 1990), and endothelin-3 (Matsumoto et al. 1989; Shinmi et al. 1989b). Three observations support the notion that these endothelins are produced locally:

(i) mRNA for endothelin-1 is present in regions of the brain which also contain the "mature", physiologically-active form of the polypeptide (MacCumber et al. 1989; Yoshizawa et al. 1990);

(ii) "big-endothelin", the immediate precursor of the "mature" twenty-one amino acid endothelin-1 (Chapter 3), together with the C terminal fragment of "big-endothelin" (residues *23–39*), are found together with the mature endothelin in the same regions of the brain (Shinmi et al. 1989b); and

(iii) the hypothalamus contains mRNA encoded for endothelin-3 (Bloch et al. 1989a, b).

Endothelin-1 and Endothelin-3 Distribution in the Brain

The endothelins are distributed throughout the brain in a non-homogeneous manner. In rat brain, for example, the cerebrum, cerebellum, medulla oblongata and pituitary gland all contain their own reservoirs of endothelin-1 and endothelin-3 (Table 12.1). Particularly high levels of endothelin-1 are present in the cerebral cortex, levels of 160 fmol per gram of tissue not being unusual (Matsumoto et al. 1989). The cerebellum, medulla oblongata and pituitary gland also contain relatively high levels of this polypeptide, the rank order of concentration being cerebral cortex > cerebellum > medulla oblongata > pituitary gland (Table 12.1).

Table 12.2. Endothelin-1 (ET-1) and endothelin-3 (ET-3) content of homogenized porcine brain and spinal cord

Tissue	fmol per gram tissue		
	ET-1	ET-3	ET-1/ET-3
Brain	140	5	28
Spinal cord	120	9	13

Data from Shinmi et al. 1989b

As far as endothelin-3 is concerned it's highest concentration in the brain (around 72 fmol per gram tissue) seems to occur in the pituitary gland (Table 12.1), where its concentration actually exceeds that of endothelin-1. Hence, whereas for endothelin-1 the concentration in the cerebral cortex > cerebellum > medulla oblongata > pituitary gland, a different rank order exists for endothelin-3, its concentration in the pituitary gland > cerebellum > cerebral cortex > medulla oblongata. The pituitary gland seems to be the only place where the concentration of endothelin-3 actually exceeds that of endothelin-1. Even in the intestine and lungs, where both isoforms are present in quite high concentrations (Table 12.1), endothelin-1 is the predominant isoform. The same holds true for the spinal cord, where (Table 12.2) the concentration of endothelin-1 > endothelin-3.

The question naturally arises as to whether the endothelins which are present in brain tissue have a functional role under normal conditions? It is almost impossible to answer this question at the moment, because the necessary experiments have not been undertaken. However, as far as endothelin-1 is concerned, its level in the brain is known to fall during water deprivation (Yoshizawa et al. 1990), indicating that it may be released under such conditions. Maybe this provides a way of modulating the local release of the neurosecretory hormones, vasopressin and oxytocin.

The presence of relatively large amounts of endothelin-3 in the intestine and pituitary gland, when coupled with its presence in specialized regions of the brain, is reminiscent of the situation which applies to other biologically active polypeptides – including gastrin, vasoactive intestinal polypeptide, neuropeptide Y and cholecystokinin. Maybe this resemblance is purely co-incidental – or it could be, as others have suggested (Matsumoto et al. 1989), that endothelin-3 really belongs to the family of brain-gut peptides. Whether this is true or not is immaterial for the moment. The important points to grasp are:

(i) *the endothelins (endothelin-1 and endothelin-3) are produced in the brain;*

(ii) *their distribution is non-homogeneous, with high levels of endothelin-1 being found in the cerebral cortex, and relatively high levels of endothelin-3 in the pituitary gland; and*

(iii) *these polypeptides may function as neuropeptides.*

The Occurrence and Distribution of Specific high Affinity Binding Sites for Endothelin (Endothelin-1) in the Brain

Binding sites for [^{125}I]endothelin-1 have been identified in rat and human brain by *in vitro* autoradiography (Koseki et al. 1989a, b; Jones et al. 1989a; Hoyer et al. 1989; Kohzuki personal communication). The distribution of these binding sites display a characteristic pattern quite distinct from other neuropeptides. Areas of high density

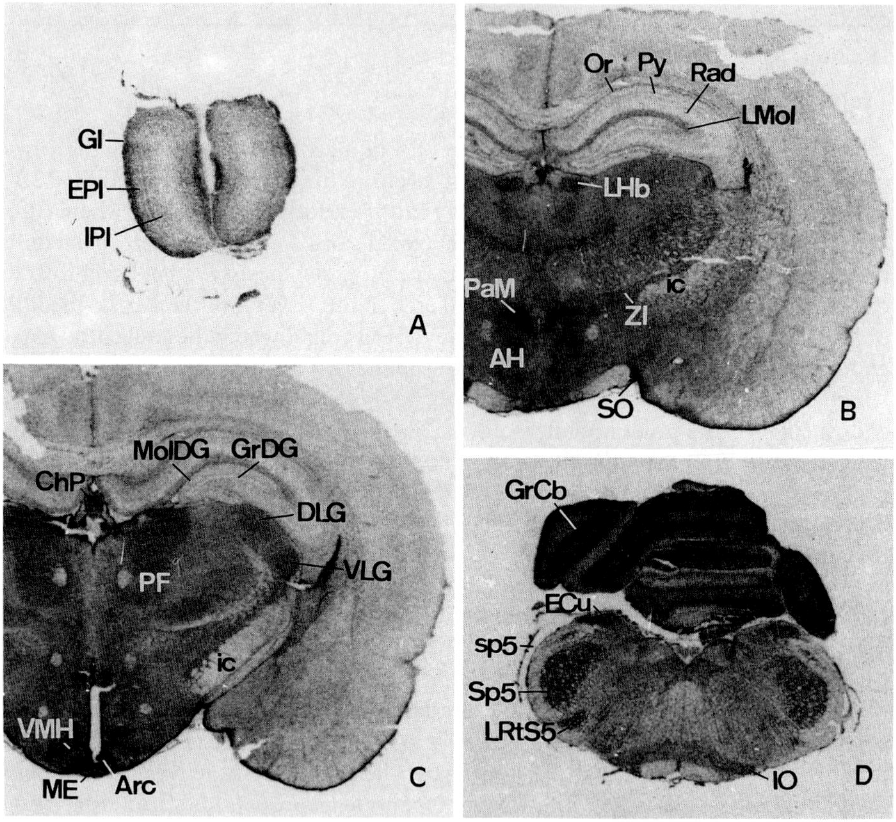

Fig. 12.1. Autoradiographs showing [^{125}I]endothelin binding to coronal sections through a rat brain at the level of the (A) olfactory bulb, (B) mid-hypothalamic region, (C) caudal hypothalamic region and (D) medulla oblongata. The abbreviations used are (A) EPl: external plexiform layer, Gl: glomerulosa layer, IPl: internal plexiform layer; (B) AH: anterior hypothalamic area, ic: internal capsule, LHb: lateral habenula, LMol: lacunosum moleculare layer of the hippocampus, Or: oriens layer of the hippocampus, PaM: magnocellular part of the paraventricular hypothalamic nucleus, Rad: radiatum layer of the hippocampus, SO: supraoptic hypothalamic nucleus, ZI: zona incerta; (C) Arc: arcuate nucleus, ChP: choroid plexus, DLG: dorsal lateral geniculate nucleus, GrDG: granular layer of the dentate gyrus, ic: internal capsule, ME: median eminence, MolDG: molecular layer of the dentate gyrus, PF: parafascicular nucleus of the thalamus, VLG: ventral lateral geniculate nucleus; (D) ECu: external cuneate nucleus, GrCb: granular layer of the cerebellum, IO: inferior olivary nucleus, LRtS5: lateral reticular nucleus – subtrigeminal part, sp5: spinal trigeminal tract, Sp5: spinal trigeminal nucleus. (This figure was kindly supplied by Drs. M. Kohzuki and S-Y Chai)

Table 12.3. Heterogeneous distribution of endothelin-1 binding sites in rat brain

Tissue	Density (fmol per mg tissue)
Hippocampus, CAI	372
Thalamus	517
Hypothalamus	727
Cerebellum	882
Olfactory bulb	493
Superior colliculus	741
Caudate putamen	255

Data from Hoyer et al. 1989
This list is not meant to be exhaustive. It is included here to illustrate the non-homogeneous distribution of endothelin-1 binding sites in the brain

binding include the glomerular layer of the olfactory bulb (Figure 12.1 A), choroid plexus (Figure 12.1 B, C), many hypothalamic structures including the paraventricular (Figure 12.1 B), supraoptic (Figure 12.1 B), ventromedial (Figure 12.1 C) and arcuate nuclei (Figure 12.1 C), median eminence (Figure 12.1 C) and the cerebellum (Figure 12.1 D). Other regions which display moderate to high binding site densities include many thalamic, and the vestibular, reticular and trigeminal nuclei. The region of highest density is the cerebellum (Table 12.3).

As far as the characteristics of these binding sites are concerned they resemble the endothelin-1 binding sites found in other tissues (Chapter 7). They are of high affinity, and binding is saturable, with an apparent affinity constant of around 3 nM, and a capacity (B_{max}) of about 200 fmol per mg protein. Binding is relatively slow, and dissociation even slower so that less than ten percent of the specifically bound endothelin dissociates over a time period of forty eight hours (Jones et al. 1989a). The binding is *not* inhibited by angiotensin II, vasopressin, parathyroid hormone nor atrionatriuretic peptide, nor is it inhibited by calcium antagonists. Presumably the endothelin binding sites in the brain (hypothalamus, cortex, cerebellum and caudate putamen) also recognize the sarafotoxin SRTX-b toxin of the asp venom (Chapter 4) because this toxin competes with endothelin-1 for the binding sites in brain membranes (Ambar et al. 1989a).

Interestingly, ET-1, ET-2 and ET-3 all display quite similar affinities when competing for [^{125}I]endothelin-1 binding sites in rat cerebellar sections. This suggests that the receptor sites in the cerebellum recognise and bind all three forms of endothelin (Kohzuki personal communications) with similar potencies, despite the fact that the concentration of ET-1 usually exceeds that of ET-3 – as can be seen in Table 12.1.

What is the Functional Significance of Endothelin Binding in the Brain?

As yet uncertainty exists as to whether the centrally produced endothelins are functionally significant, but the fact that the polypeptides are produced locally, and in areas of the brain which are well endowed with endothelin-specific receptors implies that they are of some importance. The detection of endothelin-like immunoreactivity and endothelin mRNA in the paraventricular nucleus, and the presence of high densities of binding sites in this structure and in the arcuate nucleus and median eminence suggests that these particular polypeptides may contribute to the processes which normally regulate hypothalamic and pituitary hormone release. In other words they may function as central regulators of water homeostasis, in addition to their possible central effects on blood pressure control. In the case of the pituitary hormone, vasopressin (VP), the possibility of the endothelins being natural modulators of its release warrants careful consideration, because:

(i) the genes encoding for vasopressin are expressed in the hypothalamus and the mature hormone is released from the posterior (neural) lobe of the pituitary gland;
(ii) mRNA for endothelin-1 and endothelin-3, together with immunoreactive endothelin-1 and endothelin-3 and their receptors, occur in the same areas of the brain; and
(iii) under *in vitro* conditions, endothelin-1 is a potent stimulator of hypothalamic vasopressin release (Shichiri et al. 1989).

A similar argument may well apply to oxytocin.

Mode of Action

When considering how the endothelins might affect the brain it is necessary to separate the synaptosomes from the supporting astrocytic/glial cells.

Synaptosomes

Whatever their function, the cascade of subcellular events which is initiated by the interaction of the endothelins with their binding sites in brain synaptosomes resembles, in part, that described in Chapter 6 for vascular smooth muscle, and in Chapter 12 for cardiac muscle because phosphoinositide hydrolysis is stimulated (Ambar et al. 1989a). However, *in marked contrast to the events which occur in smooth and cardiac muscle cells, the endothelin-induced stimulation of phosphoinositide hydrolysis in brain synaptosomes, which are known to contain endothelin specific binding sites, is accompanied neither by a raised cytosolic Ca^{2+}, nor by the opening of non-specific cation channels, nor by an enhanced rate of Ca^{2+} efflux (Hamilton et al. 1989).* Hence the sequence of events which is triggered when the endothelin-1 receptors in brain synaptosomes are activated appears to differ from that which occurs in smooth and cardiac muscle, where cytosolic Ca^{2+} rises. One possible interpretation of this difference is that the endothelin receptors in brain synaptosomes and those in the heart and vasculature belong to different subpopulations – one population being associated with stimulation of phosphoinositol metabolism and the other with the enhanced Ca^{2+} influx. In the heart and smooth muscle cells of the vasculature it could be argued that both receptor types are present whereas brain synaptosomes would only contain the phosphoinositol-linked receptors.

Astrocytic/Glial Cells

These are enzyme-rich cells. They possess glial end-feet which directly appose the endothelial cells of the brain capillaries (forming the *glia limitans*) as well as feet which envelope and possibly provide mechanical support for, the neuronal synapses. When exposed to the endothelins and the sarafotoxins, they respond in a manner which is reminiscent of that displayed by cardiac and smooth muscle cells, because they exhibit a dose-dependent and biphasic increase in cytosolic Ca^{2+} (Supattapone et al. 1989). This biphasic rise in cytosolic Ca^{2+} is characterized by a calcium antagonist-insensitive spike followed by a calcium antagonist-sensitive but sustained plateau phase. Sarafotoxin SRTX-b has a similar effect but it is less potent than either endothelin-1 or endothelin-3, the rank order of potency in increasing cytosolic Ca^{2+} in these cells being endothelin-3 > endothelin-1 > SRTX-b (EC_{50} values of 10, 30 and 40 nM respectively).

The endothelins have another interesting effect on the astrocytic/glial cells in that they stimulate [^{3}H]thymidine uptake (Supattapone et al. 1989). Such an effect almost certainly means that these polypeptides are mitogens for the glial cells – just as they are for smooth muscle cells (Nakaki et al. 1989).

Endothelin-1 and the Cerebral Arteries

Endothelin-1 has an interesting effect on cerebral arteries. Surprisingly, even when administered in relatively large doses, the bolus injection of endothelin-1 into the vertebral artery has no appreciable effect on basilar artery caliber. By contrast, the intracisternal injection of the polypeptide triggers a dose-dependent and sustained constriction (Shigeno et al. 1989). This constrictor response occurs irrespective of whether the endothelium is intact, indicating that the response does not depend on local release of a constrictor substance (Asano et al. 1989). At the same time, the fact that when injected into the vertebral artery endothelin-1 fails to trigger a constrictor response, whereas an intracisternal injection produces a sustained constrictor response must surely be interpreted as meaning that endothelin-1 acts from the adventitial (not luminal) side of the cerebral blood vessels. Under these conditions relatively low doses of endothelin (of the order of 10 pmol), given intracisternally, cause sustained cerebral artery vasospasm (Asano et al. 1989). There could be an interesting sequence of events here, because thrombin formation and tissue injury promote mRNA endothelin-1 expression and endothelin-1 secretion (Yanagisawa et al. 1988a). Could this be why sustained cerebral vasospasm is often encountered as a delayed event after cerebral ischaemia and injury? Is it the locally produced endothelin-1 acting on the adventitial surface of the cerebral vessels which is responsible for this delayed but sustained constriction, and could such an effect be involved in the delayed cerebral vasospasm which often afflicts patients with aneurysmal subarachnoid haemorrhage? It is not too difficult to imagine this being the case, because plasma endothelin levels increase under these conditions (Masaoka et al. 1989) and even if only a small amount of the polypeptide gained access to the adventitial surface of the cerebral arteries, prolonged vasospasm could occur, without necessarily altering heart rate or blood pressure (Asano et al. 1989).

However, there is another and equally viable approach to the possible involvement of the endothelins in cerebral vasospasm, because the endothelium of the cerebral blood vessels is also capable of releasing vasodilator agents. Since removal of the endothelium potentiates the constrictor effect of endothelin-1 on cerebral arteries (Kauser et al. 1990), the constrictor response which normally develops reflects a balanced response determined by the direct constrictor effect of the polypeptide on the one hand and its capacity to stimulate the release of the endothelial-derived vasodilator substances. Cerebral vasoconstriction, therefore, could equally well involve either the direct constrictor response or the failure of the endothelium-dependent dilator effect. Gross tissue injury, or ischaemia, could easily cause this balance to be disturbed, particularly if patches of the endothelium are injured.

Summary

1. mRNA for endothelin-1 and endothelin-3 occurs in various regions of the brain, including the hypothalamus.
2. "Mature" endothelin-1 and -3 are present in relatively high concentrations in the cerebral cortex, cerebellum, medulla oblongata and pituitary gland.
3. Specific high affinity endothelin-1 binding sites occur in the hippocampus, thalamus, hypothalamus, cerebellum, olfactory bulb, superior colliculus and caudate putamen.
4. When added to isolated synaptosomes endothelin-1 stimulates phosphoinositol metabolism, *without* increasing cytosolic Ca^{2+}.
5. When added to astrocytic glial cells, the endothelins increase cytosolic Ca^{2+} (ET-3 > ET-1) and stimulate [^{3}H]thymidine uptake, indicating a possible mitogenic effect.
6. The co-distribution of the endothelins and their binding sites with cells which are responsible for producing and secreting vasopressin, coupled with the ability of endothelin-1 to stimulate vasopressin release, may mean that these peptides function as neuropeptides. They could be involved in the central mechanisms which are responsible for maintaining not only blood pressure but also volume homeostasis.
7. As far as the cerebral vasculature is concerned, the endothelins evoke a sustained constrictor response which is endothelium-independent but which becomes exaggerated if the endothelium is removed. Each constrictor response, therefore, must represent a balance between the direct constrictor response and dilation caused by the release of endothelium-derived relaxing agents.

8. Under *in vivo* conditions an endothelin-induced pressor response may involve its central pressor effect, in addition to its direct constrictor effect on the smooth muscle cells of the vasculature.
9. Under certain pathological conditions, locally released endothelins may trigger sustained cerebral vasospasm – such as that which often occurs in a delayed response *after* subarachnoid haemorrhage, or cerebral ischaemia.

Chapter 13
Endothelin and the Coronary Circulation

"Work on, my medicine, work! Thus credulous fools are caught".
WILLIAM SHAKESPEARE: Othello

It is reasonable to assume that Cleopatra's death was due to coronary insufficiency caused by the sarafotoxins (primarily sarafotoxin SRTX-b) in the asp's venom (Chapter 4). The endothelins are also potent coronary constrictors. In anaesthetized dogs, for example, a bolus injection into the coronary circulation of as little as 30 pmol of endothelin-1 per kilogram body weight reduces coronary blood flow by more than ninety percent. Sometimes this dose even causes total coronary occlusion (Kurihara et al. 1989b). Large epicardial as well as small resistance vessels are affected, and the effect lacks species specificity, human (Chester et al. 1989), rat (Fukuda et al. 1989), pig (Yanagisawa et al. 1988a), dog (Clozel and Clozel 1989; Kurihara et al. 1989a; Cocks et al. 1989) and rabbit (Karwatowska-Prokopczuk and Wennmalm 1990) coronary arteries, and in some cases veins (Cocks et al. 1989) all having been shown to be sensitive.

The Coronary Constrictor Effect of Endothelin-1

The coronary constrictor effect of endothelin-1 has been studied in considerable detail, the constrictor response often being contrasted with that of the synthetic calcium channel agonist Bay K8644 (Table 13.1). The endothelin response can be characterized as follows:

Table 13.1. Comparison between the coronary vasoconstrictor properties of Bay K8644 and endothelin-1

Response	Bay K8644	Endothelin-1	Reference
Rate of onset of constrictor response	Fast	Slow	Stasch and Kazda 1989
Rate of recovery on washout	Fast	Slow	Baydoun et al. 1989
Dependence on membrane depolarization	Yes	No	Stasch and Kazda 1989
Nicardipine-induced inhibition	Competitive	Non-competitive	Stasch and Kazda 1989

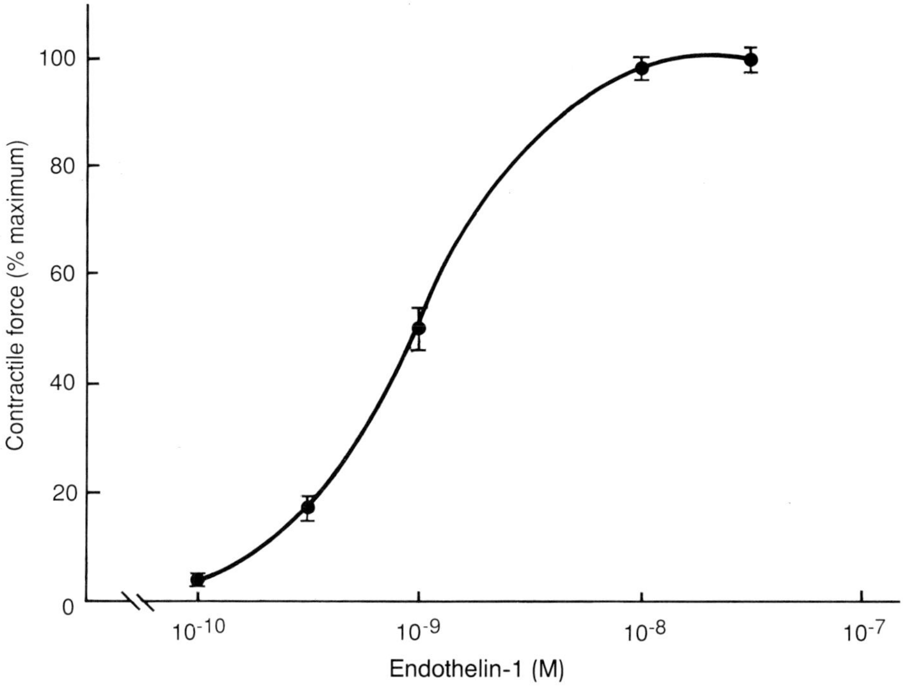

Fig. 13.1. Dose-dependent constrictor effect of endothelin-1 on human left anterior descending coronary arteries. (Data from Chester et al. 1989)

(i) it is dose-dependent, and lacks species specificity (Figure 13.1, and Table 13.2);
(ii) the response develops slowly and is sustained for long periods of time – sometimes for as long as sixty minutes after the polypeptide has been removed (Clozel and Clozel 1989);
(iii) as with the constrictor effect of endothelin-1 on other vascular beds, its effect on the coronary vasculature is insensitive to:
  (a) α adrenergic blockade;
  (b) serotonergic blockade;
  (c) histaminergic blockade;
  (d) cyclooxygenase inhibitors, and
  (e) angiotensin converting enzyme (ACE) inhibitors.

In addition,
(iv) coronary veins, as well as coronary arteries are affected, veins being more sensitive than arteries (Figure 13.2, and Cocks et al. 1989);
(v) large subepicardial are more sensitive than subendocardial coronary arteries (Clozel and Clozel 1989);
(vi) the response does *not* require an intact endothelium, and therefore cannot be attributed to the release of endothelial-derived vasoconstrictor agents, including thromboxane A_2;

Table 13.2. Lack of species specificity of the coronary constrictor effect of endothelin-1

Species	Reference
In Vitro Studies	
A. Coronary artery rings of	
(i) Swine	Yanagisawa et al. 1988a
	Tippins et al. 1989
	Stasch and Kazda 1989
	Power et al. 1989a, b
	Kasuya et al. 1989a, b
(ii) Dog	Cocks et al. 1989
(iii) Human	Chester et al. 1989
B. Isolated perfused hearts	
(i) Rats	Fukuda et al. 1989
	Baydoun et al. 1989a
(ii) Rabbit	Karwatowska-Prokopczuk and Wennmalm 1990
In Vivo Studies	
Anaesthetized dogs	Kurihara et al. 1989
	Clozel and Clozel 1989
Pithed rats	Mir et al. 1989

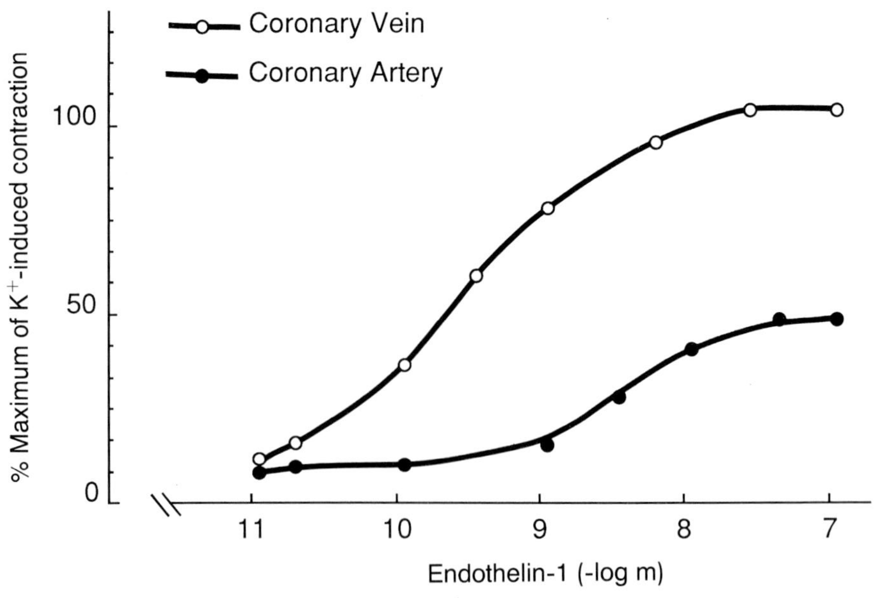

Fig. 13.2. Concentration-contraction curves to endothelin-1 in coronary artery and vein rings from greyhound dogs. Note the relative hypersensitivity of the venous preparations. (Data from Cocks et al. 1989)

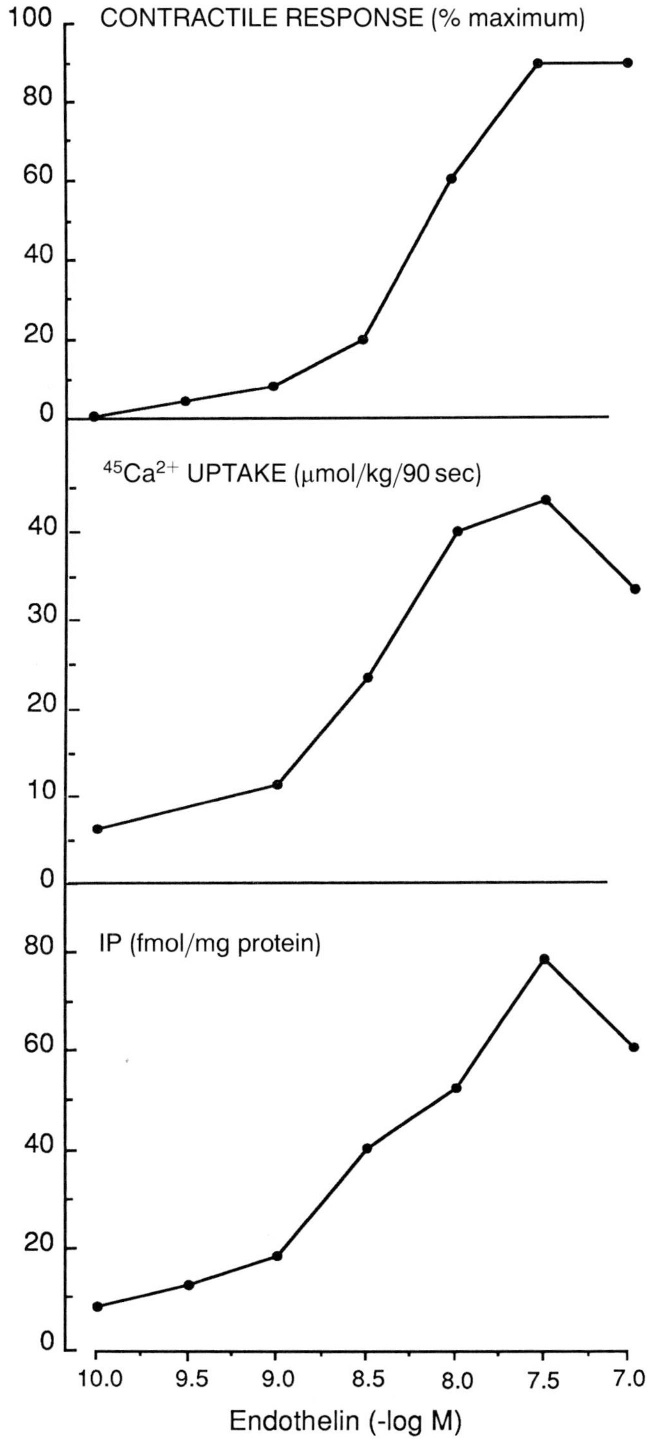

Fig. 13.3. Concentration-dependent effects of endothelin-1 on isometric force (A), rate of $^{45}Ca^{2+}$ uptake (B) and inositol monophosphate (IP) accumulation (C) in dog coronary artery rings. Data from Pang et al. (1989)

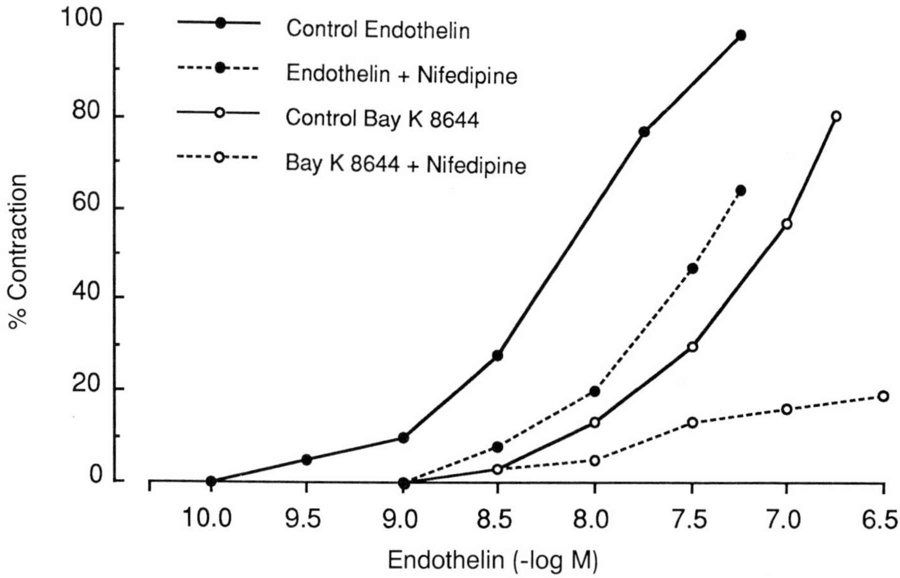

Fig. 13.4. Comparison between the coronary constrictor effect of endothelin-1 and Bay K8644 on porcine coronary artery rings. Bay K8644 is a synthetic Ca^{2+} agonist which directly activates the Ca^{2+} channels

(vii) small resistance vessels are more sensitive than the larger capacitance vessels (Tippins et al. 1989), and
(viii) many vasodilators, including
  (a) isoprenaline,
  (b) substance P,
  (c) adenosine, and
  (d) glyceryl trinitrate relax endothelin-1 constricted arteries (Chester et al. 1989).

In addition,
(ix) the constrictor effect is dependent upon the presence of extracellular Ca^{2+} (Yanagisawa et al. 1988a; Kasuya et al. 1989a; Baydoun et al. 1989);
(x) it is accompanied by an accelerated rate of Ca^{2+} uptake (Pang et al. 1989; Kasuya et al. 1989a, b, and Figure 13.3) and enhanced rate of phosphoinositol metabolism;
(xi) a wide variety of calcium antagonists – including verapamil, nicardipine, and nitrendipine (Yanagisawa et al. 1988a; Baydoun et al. 1989; Chester et al. 1989; Stasch and Kazda 1989) attenuate the coronary constrictor response but in a *non-competitive* manner. This means (Figure 13.4) that the calcium antagonists do not simply shift the dose-response curve for the constrictor effect of endothelin-1 to the right, as happens with competitive antagonism. Instead the maximum response to endothelin-1 is reduced, indicative of non-competitive inhibition.

(xii) As in the peripheral vasculature, the coronary constrictor effect of endothelin-1 is often preceded by a *transitory* dilator response (Cocks et al. 1989).

The Biochemical Basis of the Coronary Constrictor Effect of Endothelin-1

Is the Endothelium Involved?

There are two main reasons for rejecting this hypothesis.

Firstly, perfusion of air through the coronary circulation, a process which is often used to remove the endothelial lining from the coronary vasculature, does not attenuate but rather *potentiates* the constrictor effect of the polypeptide (Baydoun et al. 1989).

Secondly, the involvement of other endothelial-derived vasoconstrictor agents (such as the thromboxanes) has been excluded, because

(i) endothelin-1 induced coronary vasoconstriction appears *not* to be accompanied by thomboxane release, and

(ii) inhibitors of thromboxane release (fluriboprofen and SW755G) do not attenuate the response (Baydoun et al. 1989).

Is the Response Due, In Part at Least, To the Activation of the Sympathetic Nervous System With the Concommitant Release of Noradrenaline?

This possibility can also be discarded, because the response is not modified by α adrenergic blockade (Chester et al. 1989).

Is the Platelet Activating Factor (PAF) Involved?

The prolonged vasoconstrictor effect of endothelin-1 on the coronary vasculature bears some resemblance to that of PAF, which also causes a dose-dependent and sustained constriction. The involvement of PAF in the coronary constrictor effect of endothelin-1 can be excluded, however, because WEB 2086, a specific PAF antagonist, does not alter the coronary constrictor effect of endothelin-1 (Baydoun et al. 1989).

Does the Response Involve Increased Calcium Ion Influx Through the Slow Calcium Channels?

Since the coronary constrictor effect of endothelin-1 is accompanied by an enhanced rate of Ca^{2+} influx (Pang et al. 1989), and it is attenuated or abolished by pretreatment with calcium antagonists (Yanagisawa et al. 1988a; Kasuya et al. 1989a) it is reasonable to assume that the coronary constrictor effect of the polypeptide involves activation of the slow Ca^{2+} channels.

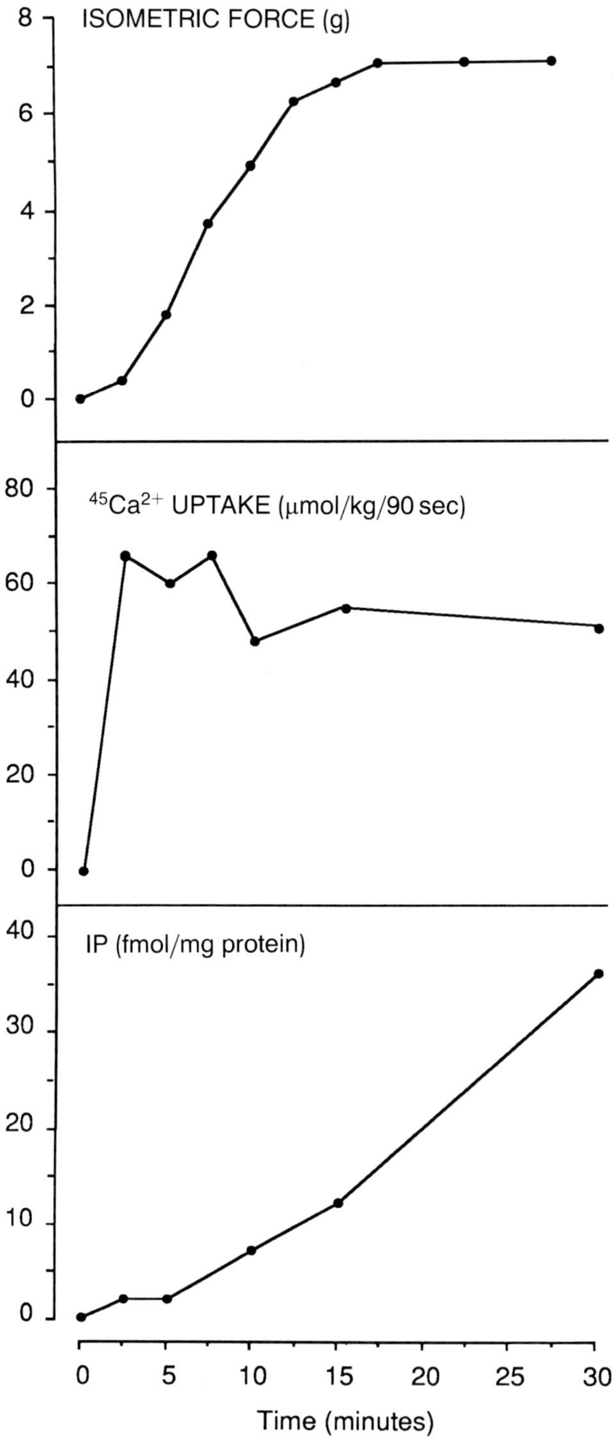

Fig. 13.5. Time course for the effect of 3×10^{-8}M endothelin-1 (30 nM) on (A) isometric force development, (B) $^{45}Ca^{2+}$ uptake and (C) inositol phosphate accumulation in canine coronary artery rings. (Data from Pang et al. 1989)

As can be seen from Figure 13.4, however, the inhibitory effect of the calcium antagonists bears all the hall marks of non-competitive antagonism, with the dose-response curves being moved to the right, but with the maximum response being reduced. Such a response is quite different from that obtained when the calcium agonist Bay K8644 is used to contract the coronary vasculature, since the calcium antagonists displaces the dose response curve for the constrictor effect of this substance to the right, but *without* attenuating the maximum response. Hence it seems that although endothelin-1 and Bay K8644 both constrict coronary blood vessels, and although in each case the response is Ca^{2+}-antagonist sensitive the intermediate events are different. In the case of Bay K8644, the substance has a direct excitatory effect on the channels, an effect which is triggered by its direct interaction with the dihydropyridine binding sites in the channel complex (Schramm et al. 1983). Endothelin-1 does *not* interact directly with any of the known calcium channel binding sites (Kasuya et al. 1989a). Probably therefore, by interacting with its own receptors, it indirectly activates these channels, with a resultant increase in Ca^{2+} influx.

It is difficult to escape the conclusion that the coronary constrictor effect of endothelin-1 involves an increased influx of Ca^{2+} through the slow Ca^{2+} channels (the response being calcium antagonist-sensitive, requiring the presence of extracellular Ca^{2+}, and being accompanied by an increased inward Ca^{2+} current (Silberg et al. 1989)). However Ca^{2+} from another source may also be involved, because, as Figure 13.3 shows and other investigators have confirmed (Kasuya et al. 1989b), the endothelin-1-induced coronary constrictor response is also accompanied by a dose-dependent increase in phosphoinositol metabolism. Moreover, and as can be seen from Figure 13.5, the time course for the contractile response roughly parallels that for the increase in Ca^{2+} influx and phosphoinositol metabolism. *Perhaps the best way of interpreting these various observations is to assume that the coronary constrictor effect of endothelin-1 depends upon both an increase in Ca^{2+} influx through indirectly activated calcium channels, and an inositol triphosphate-mediated release of intracellular Ca^{2+}. From a therapeutic point of view the involvement of the calcium channels is highly significant because it means that the endothelin-induced coronary constriction is calcium antagonist sensitive.*

Endothelin-1 Receptors in the Coronary Vasculature

The coronary blood vessels contain specific binding sites for endothelin. These binding sites have been located by standard radioautography using iodinated endothelin-1 ($[^{125}I]$ET-1) as the ligand. The binding sites are:

(i) present in the walls of arteries and veins (Power et al. 1989b);
(ii) absent from the endothelium (Power et al. 1989a, b; Chester et al. 1989), and
(iii) present in the accompanying nerve trunks (Power et al. 1989a, b). Also,
(iv) the binding occurs with a density (B_{max}) of approximately 0.2 fmol/mm^2 and an affinity (K_D) of approximately 0.14 nM (Power et al. 1989a, b);
(v) the binding is to a single population of sites and occurs in epicardial and endocardial vessels alike;

Table 13.3. Relative density of specific endothelin-1 binding sites in coronary arteries

Species	Tissue	[^{125}I]endothelin-1 binding site density (fmol/mm^2)	Reference
Rat	Aorta	0.22	Davenport et al. 1989b
	Pulmonary artery	0.25	Davenport et al. 1989b
	Left atria	5.9	Davenport et al. 1989b
Pig	Coronary artery	0.2	Power et al. 1989a, b
	Coronary artery	0.25	Davenport et al. 1989b
	Left ventricle	0.28	Davenport et al. 1989b
	Right atria	0.43	Davenport et al. 1989b

(vi) the density of binding in the coronary arteries (Table 13.3) is approximately the same as that found in other vessels, including the aorta and pulmonary artery. Nor does the density of binding exceed that of the heart;
(vii) the specifically bound endothelin is *not* displaceable by calcium antagonists or by a wide variety of other vasoactive polypeptides, including angiotensin II, vasopressin and neuropeptide Y.

In general therefore, the endothelin-1 binding sites in the coronary vasculature are just like those in other tissues, including the myocardium, the peripheral vasculature, the brain, the kidney, etc (Chapter 7).

The Relative Sensitivity of Coronary Arteries and Veins to the Constrictor Effect of Endothelin-1

Mention has already been made of the fact that the constrictor effect of endothelin-1 on coronary arteries is Ca^{2+}-dependent, dose-dependent and sustained. The coronary constrictor effect of the polypeptide exceeds that of other constrictors – including Bay K8644, histamine, acetylcholine, angiotensin II and prostaglandin $F_{2\alpha}$ (Kasuya et al. 1989a; Yanagisawa et al. 1988a). In addition, and as can be seen from Table 13.4, coronary arteries are no more sensitive than other arteries, but veins are more sensitive than arteries. The fact that coronary arteries are no more sensitive than other arteries might have been anticipated, because they all contain approximately the same number of receptors (Table 13.3). The relative hypersensitivity of coronary veins means that coronary veins follow the general pattern (de Nucci et al. 1988) of veins being more sensitive than arteries. As yet it is not known why veins exhibit this hypersensitivity, but, from the practical point of view it may mean that the polypeptide is involved in maintaining venous tone – which could be important in certain hypotensive crises, including shock.

As far as coronary arteries are concerned large epicardial as well as small resistance vessels respond to endothelin-1, but it seems that the small vessels are

Table 13.4. EC_{50} values for the constrictor effect of endothelin-1 on coronary and other blood vessels

Tissue	EC_{50} (nM)	
	Artery	Vein
Coronary	8.7	
Femoral	8.6	
Internal mammary	8.3	
Mesenteric	7.9	8.6

Data taken from Cocks et al. 1989. Arteries and veins were from greyhound dogs
EC_{50} values: the negative logarithim of the concentration of endothelin-1 required to produce 50 percent of the maximal response

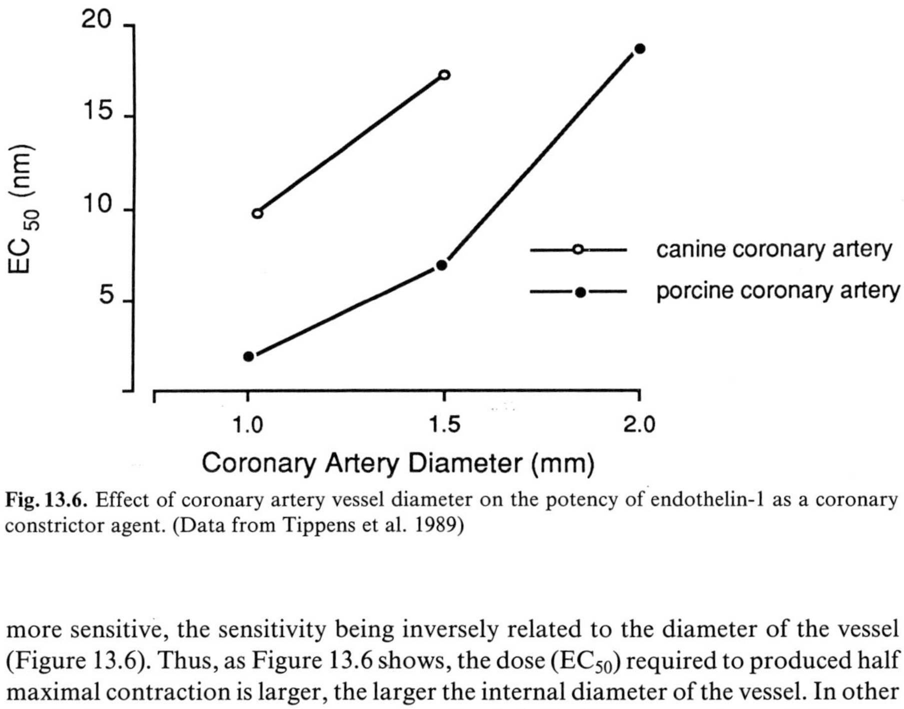

Fig. 13.6. Effect of coronary artery vessel diameter on the potency of endothelin-1 as a coronary constrictor agent. (Data from Tippens et al. 1989)

more sensitive, the sensitivity being inversely related to the diameter of the vessel (Figure 13.6). Thus, as Figure 13.6 shows, the dose (EC_{50}) required to produced half maximal contraction is larger, the larger the internal diameter of the vessel. In other words *the coronary constrictor potency of endothelin-1 is greatest in small* diameter coronary arteries.

The Transient Endothelin-1 Coronary Vasodilation

When injected intracoronarily endothelin-1 often causes an initial transient relaxation. This relaxation is endothelium-dependent, and therefore can almost certainly be accounted for in terms of the release of an endothelial-derived relaxing substance (Karwatowska-Prokopczuk and Wennmalm 1990). However, since the most pronounced and sustained response is that of vasoconstriction, the physiological importance of the endothelin-1 induced transient relaxation is of questionable importance *unless* it is normally a continuing, on-going *event*. In other words, under non pathological conditions, the ability of endothelin-1 to *induce a secondary relaxation of the coronary vasculature may be sufficient to counteract its intrinsic and primary constrictor effect. Under pathological conditions, however, or in the presence of abnormally high levels of endothelin-1, this balance may be lost, to unmask the potent constrictor effect of the polypeptide.*

The Effect of Endothelin-1 Induced Coronary Vasoconstriction on Myocardial Function

The intracoronary injection of endothelin-1 has a marked effect on the heart. Thus:
(i) end-diastolic pressure rises;
(ii) segmental shortening declines,
(iii) the heart rate slows.

In addition,
(iv) electrocardiographic signs of myocardial ischaemia develop, and

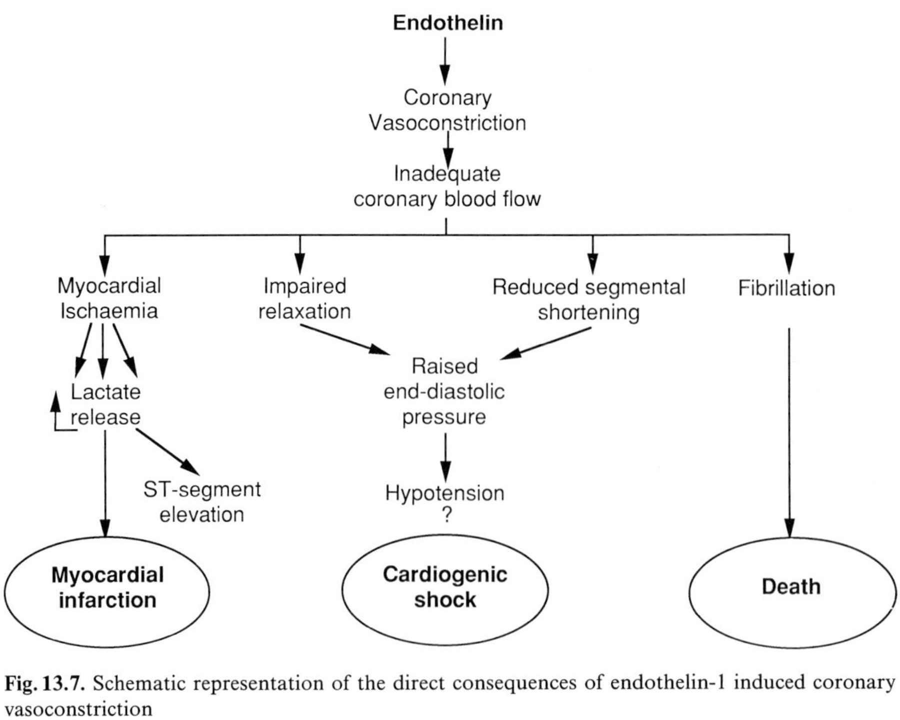

Fig. 13.7. Schematic representation of the direct consequences of endothelin-1 induced coronary vasoconstriction

(v) lactate is released from the heart (Ezra et al. 1989), indicating a switch to anaerobic metabolism, and
(vi) the glycogen, adenosine triphosphate and creatine phosphate reserves are diminished (Linz et al., 1989)

The continued intracoronary administration of endothelin-1 can have serious consequences (Figure 13.7) including severe hypotension, ventricular failure, ventricular fibrillation and death (Ezra et al. 1989; Linz et al., 1989). The hypotension is undoubtedly a secondary phenomenon caused by reduction in segmented shortening which, in turn, must be a secondary consequence of the reduction in coronary flow, since endothelin-1 is intrinsically positively inotropic (Chapter 14).

Is Endothelin-1 Involved in the Development of Coronary Vasospasm?

The answer to this question is as yet, uncertain, although the possibility has been raised on many occasions. The difficulty with the hypothesis lies in the *sustained* nature of the endothelin-induced constrictor response. Thus, whereas the endothelin-1 induced coronary artery constriction persists for up to sixty minutes, coronary artery spasm is a relatively short-lived event. This does not necessarily mean that endothelin-1 is not involved in the aetiology of coronary artery spasm however. It may contribute to the response by provoking a sustained – or background – increase in vascular tone which *may* predispose the arteries to spasm. Alternatively the spasm may reflect the temporary failure of the endothelium to release vasodilator agents, including prostacyclin and EDRF, which normally counteract the constriction caused by the endothelins.

Summary

1. Endothelin-1 is a potent coronary constrictor. Large and small coronary arteries, as well as veins, are affected.
2. The constrictor response lacks species specificity, is sustained, and is dose-dependent.
3. The potency of endothelin-1 as a coronary constrictor agent exceeds that of angiotensin II or histamine, and is inversely related to the diameter of the vessels.
4. The response is insensitive to α adrenergic and serotonergic blockade, and to inhibitors of the angiotensin converting enzyme and the cyclooxygenases.
5. Calcium antagonists (nicardipine, nifedipine) inhibit the response in a dose-dependent manner.
6. Coronary arteries contain specific endothelin-1 binding sites.
7. The constrictor response is accompanied by a dose and time-dependent increase in Ca^{2+} influx, phosphoinositol metabolism and slow inward Ca^{2+} current.
8. The intracoronary injection of endothelin-1 can cause myocardial ischaemia, hypotension and ventricular fibrillation, even at relatively low dose levels.

Chapter 14

Endothelin-1 and the Heart

"And dost thou ask why heaves thy heart, with tighten'd pressure in thy breast? Why the dull ache will not depart, by which thy life-pulse is op-press'd."
Faust – by J. W. von GOETHE

The previous chapter (Chapter 13) described the effect of endothelin-1 on the coronary circulation, where the polypeptide evokes a small but transient dilator response followed by a dose-dependent, sustained and potent constriction. The constrictor response was described as being receptor mediated, and involving

(i) an enhanced influx of Ca^{2+} ions through the Ca^{2+}-selective channels, and
(ii) a phosphoinositol-mediated mobilization of intracellular Ca^{2+}.

As far as the myocardium is concerned the endothelins have a slightly more complex effect, because under *in vivo* conditions the endothelin-induced reduction in coronary blood flow described in the last chapter leads inevitably to a reduced contractility, a slowed heart rate, a reduction in cardiac output, and electrocardiographic changes resembling those encountered during myocardial ischaemia. These ischaemia-like effects of endothelin-1 are particularly pronounced at relatively high dose levels but they are *secondary to the endothelin-induced reduction in coronary flow*. When added to isolated atrial, papillary and trabeculae muscle preparations, the secondary changes associated with the reduction in myocardial perfusion are avoided. As a result, the positive inotropic effect of the polypeptide is readily identifiable (Table 14.1, and Figure 14.1).

The threshold for the inotropic effect of endothelin-1 on atrial muscle is around 0.1 nM (Ishikawa et al. 1988a), whilst the ED_{50} value (the half-maximally effective

Table 14.1. Evidence of the positive inotropic effect of endothelin-1

Preparation	Species	Response	Reference
Atrial preparations	Guinea Pig	Positive	Ishikawa et al. 1988a
	Rat	Positive	Hu et al. 1988a
	Rat	Positive	Moravec et al. 1989
	Human	Positive	Davenport et al. 1989a
	Human	Positive	Moravec et al. 1989
Ventricular Papillary/ Trabeculae Preparations	Ferret	Positive	Shah et al. 1989
	Rat	Positive	Moravec et al. 1989
	Human	Positive	Moravec et al. 1989
	Rabbit	Positive	Watanabe et al. 1989a

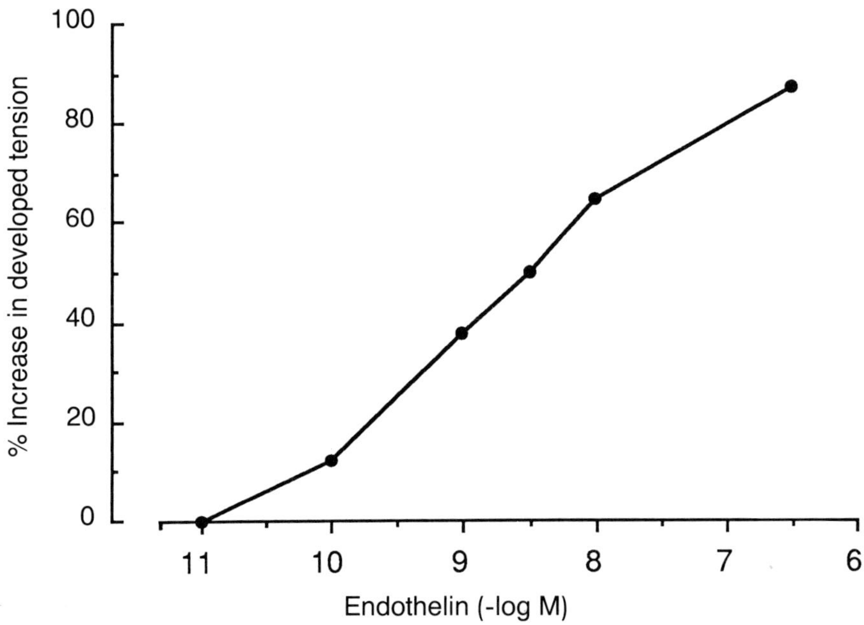

Fig. 14.1. Dose-response curve for the positive inotropic effect of endothelin-1 on electrically-stimulated guinea pig atrial preparations. Note that the inotropic effect is dose-dependent

dose) is in the low nanomolar range (Ishikawa et al. 1988a; Hu et al. 1988a). Some idea of the potency endothelin-1 has as a positive inotrope can be gauged by comparing its effect with that of the Ca^{2+} agonist Bay K8644. Thus, when added to electrically stimulated rabbit papillary muscles 1 nM endothelin-1 increases peak developed tension by more than one hundred and fifty percent. About five hundred times as much Bay K8644 (500 nM) is needed to produce a comparable increase in developed tension (Watanabe et al. 1989)!

The Characteristics of the Endothelin-1 Induced Positive Inotropic Response

The main characteristics of the positive inotropic effect of endothelin-1 have already been referred to in Chapter 6. In many ways the response resembles that obtained for the coronary vasculature. For example,

(i) the response develops slowly, taking as long as fifteen minutes to reach asymptote, and is sustained – for up to sixty or ninety minutes after washout (Moravec et al. 1989). The sustained nature of the response is almost certainly due to the slow rate at which the polypeptide dissociates from its binding sites (receptors), as described in Chapter 7.

ENDOTHELIN

ELECTROPHYSIOLOGY

1. Membrane depolarization
2. Increased inward Ca^{2+} current

MECHANICS

1. Increased heart rate
2. Increased force of contraction
3. Coronary vasoconstriction

BIOCHEMISTRY

1. Phospholipase C activation
2. Increased IP_3 metabolism
3. Stimulation of Na^+/H^+ exchanger
4. Mobilization of internally stored Ca^{2+}
5. Increased diacylglycerol
6. Release of
   (i) ANP
   (ii) EDRF
   (iii) PGI_2 +PGE_2

Fig. 14.2. Schematic representation of the effect of endothelin-1 on the electrophysiology, mechanics and biochemistry of the heart

(ii) The positive inotropic effect is insensitive to α and β-adrenergic blockade, to anticholinergic agents, and to histaminergic and serotonergic antagonists (Ishikawa et al. 1988a; Hu et al. 1988a; Watanabe et al. 1989). In addition,

(iii) the dose-rcsponse curve for the inotropic effect is shifted to the right by the calcium antagonists, including nicardipine (Ishikawa et al. 1988a) and verapamil (Vigne et al. 1989), indicating some involvement of Ca^{2+} ions entering through the voltage-sensitive L-type calcium channels.

Apart from these similarities there are several other aspects of the positive inotropic effect of endothelin-1 which warrant mention.

(i) Both its slow rate of onset and its sustained nature distinguish it from the positive inotropic responses induced by many other agents, including angiotensin and isoprenaline, because in these cases the response develops rapidly, and recovery is speedy.

(ii) The response is accompanied by a prolongation of the cardiac action potential (Figure 14.2). In rabbit papillary muscle, for example, 10 nM endothelin-1 lengthens the action potential by about sixteen percent; at the same time tension development increases by one hundred and eighty percent (Watanabe 1989). This response is quite different from that which occurs when the calcium agonist Bay K8644 is used to trigger the inotropic response, because a concentration of Bay K8644 which causes a similar lengthening of the action potential increases tension development by less than forty percent. Now, if the positive inotropic effect of endothelin-1 had been due simply to enhanced Ca^{2+} influx through the voltage-sensitive Ca^{2+} channels, as happens with Bay K8644 (Schramm et al. 1983), the relationship between the lengthening of the cardiac action potential and the magnitude of the inotropic response should be the same for both inotropic agents. Clearly this does not happen!

(iii) Even the effect of endothelin-1 on the parameters of the individual contractions is of interest, because whereas isoprenaline and many other inotropic agents shorten the time to peak tension and accelerate the rate of relaxation,

Table 14.2. Endothelin-1 binding site density in atrial, ventricular and coronary arterial muscle (amol per mm^2)

Species	Tissue				
	Atria		Ventricle		Coronary Artery
	Left	Right	Left	Right	
Pig	5.7 ± 0.2	4.3 ± 0.2	2.8 ± 0.4	2.9 ± 0.3	2.5 ± 0.5
Rat	5.9 ± 0.4	5.3 ± 0.3	3.5 ± 0.3	3.5 ± 0.4	
Humans	8.0 ± 0.3	7.6 ± 0.3	6.4 ± 0.4	6.7 ± 0.4	

Data from Davenport et al. 1989b. Data is presented as mean ± S.E.M.

endothelin-1 *neither alters the time to peak tension, nor accelerates relaxation* (Ishikawa et al. 1988a). Instead, the effect of endothelin-1 is exactly the same as that obtained when extra Ca^{2+} is added. However, whereas the positive inotropic response which is caused by adding additional Ca^{2+} is rapidly reversed when the Ca^{2+} is removed, as is the positive inotropic effect of angiotensin and vasopressin, the effect of endothelin-1 persists.

(iv) In addition, although endothelin-1 evokes a positive inotropic response from atrial and ventricular muscle, atrial muscle is more sensitive than ventricular muscle (Moravec et al. 1989). This difference can probably be explained in terms of the fact that atrial muscle contains more binding sites than ventricular muscle (Table 14.2). This does not mean that ventricular muscle is relatively depleted of these binding sites. To the contrary, (and as Table 14.2 shows) ventricular muscle contains more binding sites than some smooth muscle cells, including those of the coronary vasculature.

In summary, therefore, endothelin-1 has a positive inotropic effect on cardiac muscle. The response develops slowly, is sustained and is accompanied by a prolongation of the cardiac action potential.

The Mechanisms Involved in the Positive Inotropic Effect of Endothelin-1

The positive inotropic effect of endothelin-1 is dose-dependent (Figure 14.1), and almost certainly is a receptor-mediated, Ca^{2+}-dependent event. The Ca^{2+} ions which, in the final analysis, mediate the inotropic response probably originate (Hirata et al. 1989a, e; Vigne et al. 1989) from three sources:

(i) extracellular Ca^{2+} ions which enter by way of the voltage-sensitive calcium channels;

(ii) extracellular Ca^{2+} ions which may enter in exchange for Na^+, and

(iii) intracellular stores which are mobilized in response to the enhanced rate of phosphoinositol metabolism (Galron et al. 1989; Vigne et al. 1989).

The Myocardial Binding Sites for Endothelin-1

As mentioned in Chapter 7 cultured neonatal (Hirata et al. 1989a, e; Galron et al. 1989) and adult (Gu et al. 1989a, b) cardiac myocytes contain specific, high-affinity binding sites for the endothelins. Binding is to a single population of sites, with a K_D of around 1 nM, and a density (B_{max}) of 50,000–80,000 sites per cell (Hirata et al. 1989a, e) – or between 100–300 fmol per mg membrane protein. These binding sites are sensitive to "up" and "down" regulation but they are insensitive to the classical antagonists of the L- and N-type of calcium channel (Table 14.3). Nevertheless Ca^{2+} ion entry through the L-type of channel must be involved in the inotropic response because

(i) calcium antagonists interfere with the positive inotropic response (Ishikawa et al. 1988a);
(ii) the endothelin-1-induced increase in cytosolic Ca^{2+} includes a calcium antagonist sensitive component (Hirata et al. 1989e), and
(iii) endothelin-1 prolongs the plateau phase of the cardiac action potential (Ishikawa et al. 1988a).

Table 14.3. The specificity of endothelin-1 binding in cardiac membranes

Compound	Reference
Substances which interact with cardiac endothelin-1 binding sites	
Endothelin-1	Gu et al. 1989a
Sarafotoxin (SRTX-b)>>SRTX-c)	Nayler et al. 1989a
Substances which do not interact with cardiac endothelin-1 binding	
Calcium Antagonists	Gu et al. 1989b, c
(−) D600	
(+) cis-diltiazem	
nicardipine	
isradipine	
nifedipine	
nitrendipine	
prenylamine	
lidoflazine	
flunarizine	
w conotoxin	
Calcium agonists	
Bay K8644	Gu et al. 1989b, c
Alpha-adrenoceptor Antagonists	
Prazosin	Gu et al. 1989b, c
Others	
Ergonovine	Nayler (unpublished)

The Endothelin-1 Induced Increase in Cytosolic Ca^{2+}

As described in detail in Chapter 6 the endothelin-induced increase in cytosolic Ca^{2+} can be divided into two phases – an initial rapidly-developed calcium antagonist insensitive spike phase which lasts for only a few minutes, followed by a smaller but sustained, and calcium antagonist sensitive plateau phase (Hirata et al. 1989a). This picture is very reminiscent of that described for smooth muscle cells.

The Stimulant Effect of Endothelin-1 on Phosphoinositol Metabolism in Cardiac Muscle

Endothelin-1 does stimulate phosphoinositol metabolism in cardiac muscle (Galron et al. 1989; Vigne et al. 1989). The stimulation is dose-dependent and progresses almost linearly for at least fifteen minutes. It is calcium antagonist *insensitive* and occurs at dose levels which are in the low nanomolar range and which accordingly are in the range needed to evoke a positive inotropic response.

In general therefore, the biochemical events which are involved in the positive inotropic effect of endothelin-1 on mammalian heart muscle resemble those already described in Chapter 6 for its effect on vascular smooth muscle cells. Underscoring the response is a raised cytosolic Ca^{2+} due, in part, to the mobilization of internally stored Ca^{2+} and, in part, to enhanced Ca^{2+} entry through the *indirectly*-activated calcium channels (Figure 14.2). The initial trigger for the cascade of events which culminates in the raised cytosolic Ca^{2+}, and hence the increase in developed tension, must be the association of the polypeptide with its specific, high affinity binding sites in the membrane

The Positive Chronotropic Effect of Endothelin-1

Although endothelin-1 causes a reflex bradycardia under *in vivo* conditions, when added to isolated, spontaneously beating atrial preparations it has a dose-dependent positive chronotropic effect (Ishikawa et al. 1988b). The threshold for this effect is around 0.3 nM, and it reaches a maximum at concentrations of around 30 nM. Once again the effect develops slowly and is sustained. It is probably a receptor-mediated event, since the atrioventricular node contains an abundance (Chapter 7) of high affinity binding sites for the polypeptide. Indeed the affinity of these receptors (K_D of around 50 pM) is far greater than that of the myocyte receptors, the K_D of which is in the low nanomolar range (Hirata et al. 1989e; Yamasaki et al. 1989).

Endothelin-1 and the Release of Atrionatriuretic Peptide (ANP)

The effect of endothelin-1 on the heart cannot be dismissed simply in terms of its chronotropic and inotropic activities, because it also promotes the release of the atrionatriuretic peptide (ANP) (Fukuda et al. 1988). The endothelin-induced stimulation of ANP secretion is a dose-dependent phenomenon. The threshold concentration required for stimulation is around 0.1 nM, and the half maximal stimulatory dose is only twenty times the threshold level (2.0 nM). Hence the response occurs at the dose range of endothelin which is physiologically active – in terms of its effect on heart rate, force of contraction and vascular constriction. However the effect cannot be accounted for in terms of a secondary response linked with other changes in the circulation – because endothelin-1 stimulates ANP secretion from isolated myocytes (Fukuda et al. 1988). Precisely why endothelin-1 stimulates ANP release is unknown, but there is a fair chance that it is due to the effect of endothelin-1 on cytosolic Ca^{2+}. This conclusion is based on the fact that other agents (the Ca^{2+} ionophore ionomycin, for example, and the phorbol ester tetra decanoyl-phorbol-acetate [TPA]) also exacerbate ANP release, and these agents either directly, or, in the case of TPA indirectly through protein kinase C activation, raise cytosolic Ca^{2+} (Fukuda et al. 1988).

The stimulatory effect of endothelin-1 on ANP release is not modified by α or β adrenergic blockade or by anticholinergic agents, but it is attenuated by calcium antagonists – including verapamil (Hu et al. 1988b) and nicardipine (Fukuda et al. 1988).

Since ANP is a potent vasorelaxant, and endothelin-1 is a potent vasoconstrictor, the ability of endothelin to stimulate ANP release might be regarded as an "in built" protective mechanism which is aimed at counterbalancing the intense vasoconstrictor effect of the endothelin polypeptide. This, together with the ability of endothelin-1 to stimulate prostacyclin and EDRF release (Warner et al. 1989; de Nucci et al. 1988) from the vasculature could be of particular importance in so far as maintaining adequate coronary perfusion under conditions in which either the density of the endothelin receptors has increased, their affinity has increased, the circulating levels of endothelin-1 have risen, or some other vasoconstrictive event has occurred to increase coronary resistance.

Endothelin and the Ischaemic Heart

There are now at least five reasons for suggesting that the endothelins – and particularly endothelin-1 – may be involved in the aetiology of ischaemic heart disease – quite apart from their potential involvement in the processes which result in the formation of atherosclerotic lesions. Before listing these reasons, however, it may be helpful to review the situation which develops (Karwatowska-Prokopczuk and Wennmalm 1990) when endothelin-1 is injected into the coronary circulation. Not only do electrocardiographic signs of ischaemia (ST-segment changes) develop, but in addition (Figure 14.3) lactate is released, relaxation is impaired, and arrhythmias – including fibrillation – develop. These changes are almost identical with those caused by acute coronary occlusion! Apart from this similarity the main reasons for arguing in favour of the hypothesis that the endothelins contribute to ischaemic heart disease are as follows:

(i) endothelin is a potent coronary vasoconstrictor (Chapter 13). It's release, therefore, will result in the exacerbation of any pre-existing coronary

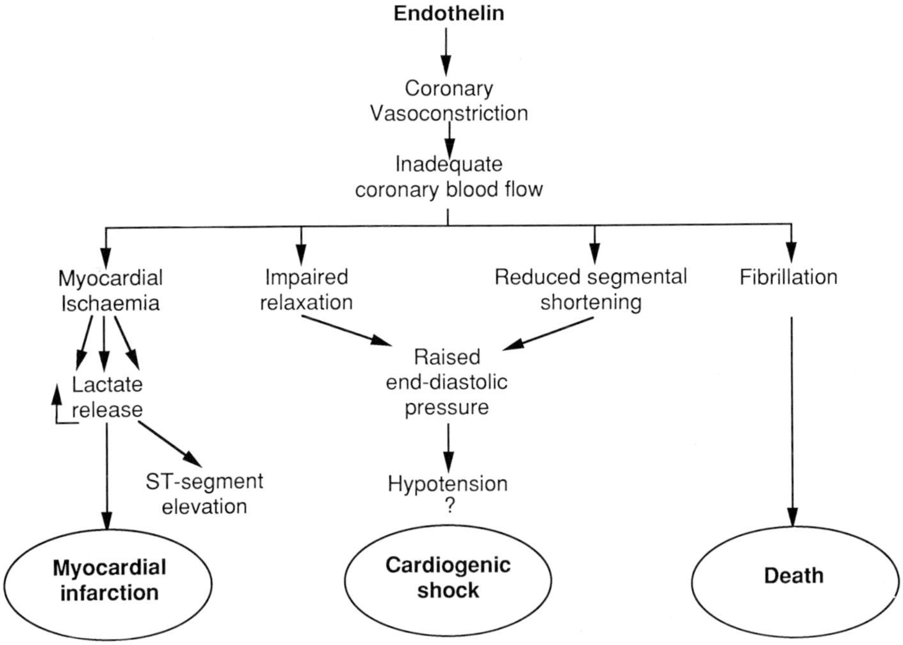

Fig. 14.3. Schematic representation of the events which occur when endothelin-1 is injected into the coronary circulation. Note that many of the resultant changes mimic those caused by ischaemia

insufficiency – irrespective of whether that constriction be caused by enhanced sympathetic tone, plaque rupture, vasospasm or some other metabolic cause.

(ii) Because of its innate ability to mobilize intracellular Ca^{2+} (Chapter 6) endothelin-1 will enhance the likelihood of cytosolic Ca^{2+} overloading occurring during the ischaemic episode. In the normally well perfused heart there is sufficient energy (as adenosine triphosphate, ATP) available to ensure that any additional Ca^{2+} which appears in the cytosol in response to endothelin (Figure 14.4) can either be retrieved by way of the sarcoplasmic reticulum or expelled across the membrane by means of the Ca^{2+}-activated ATPase. Both of these processes, however, require energy (as ATP) which, although available in the aerobically perfused hearts, is in short supply or totally depleted during an ischaemic episode. An uncontrolled rise in cytosolic Ca^{2+} is closely linked with the onset of events which results in cell death and tissue necrosis (Nayler 1983), and it seems likely that the endothelins can only exacerbate this problem!

(iii) Prolonged episodes of ischaemia increase (Liu et al. 1990a) the number of endothelin-1 binding sites which are available for interaction with newly released, or circulating endothelin. This response, which involves the externalization of the binding sites (Liu et al. 1989), is time-dependent (Figure 14.5) and, as can be seen from Figure 14.6, is exacerbated upon reperfusion. There are ways of avoiding this problem – including pretreatment with calcium

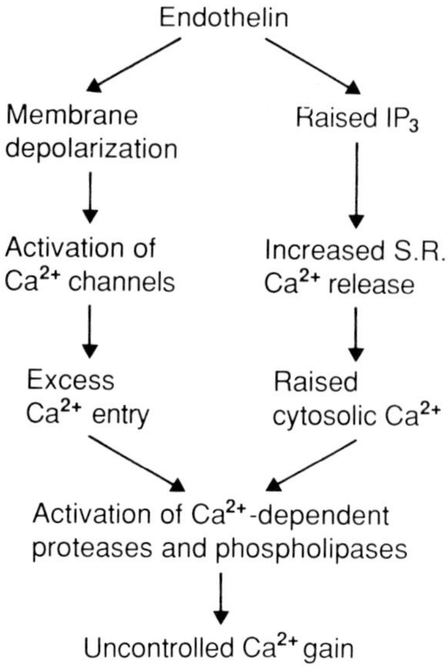

Fig. 14.4. Schematic representation of the possible involvement of endothelin-1 in the events which contribute to a raised cytosolic Ca^{2+} under conditions of adenosine-triphosphate depletion

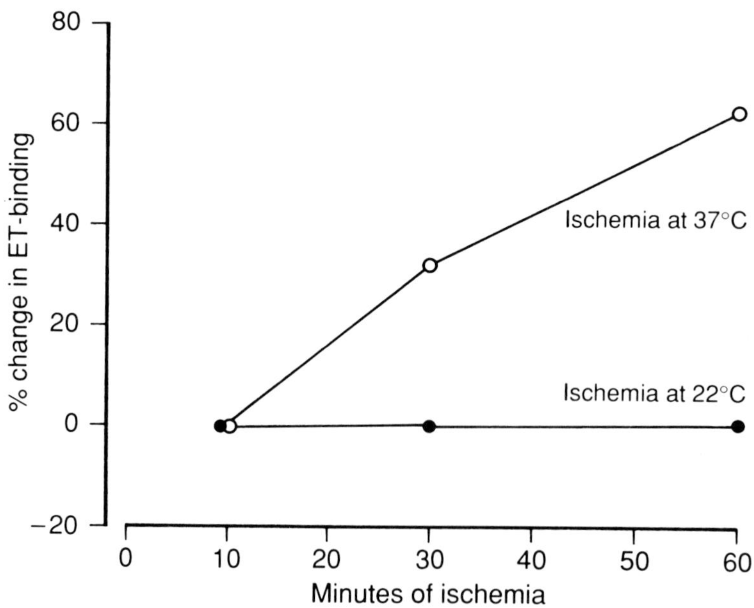

Fig. 14.5. Effect of global ischaemia at 37 °C and 22 °C on the density of endothelin-1 binding sites in isolated rat hearts. Ischaemia was for 30 or 60 minutes. Note that the ischaemia-induced increase in binding site density is abolished under hypothermic conditions

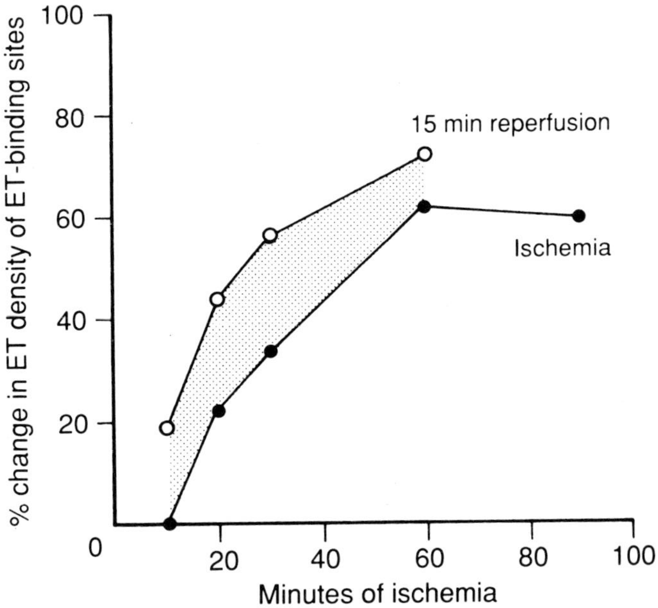

Fig. 14.6. Exacerbation of the ischaemia-induced increase in endothelin-1 binding site density in isolated rat hearts caused by *reperfusion*. The hatched area represents the difference between the binding site density caused by reperfusion. The perfusion buffer was Krebs-Henseleit solution, at 37°C

antagonists (Nayler, unpublished data), and using hypothermic conditions during the ischaemic episode (Liu et al. 1990a, and Figure 14.5).

(iv) At the same time, the circulating levels of endothelin-1 increase (Miyauchi et al. 1989) and endothelin mRNA expression is activated.

(v) Finally, and perhaps the most convincing evidence of all, is the recent demonstration by Watanabe and his colleagues (Watanabe et al. 1990) of the ability of pretreatment with a monoclonal endothelin-1 antibody to reduce experimentally-induced infarct size.

Summary

1. Endothelin-1, a natural secretagogue for the atrial natriuretic peptide, has a complex effect on the heart (Figure 14.2). It has a *direct* positive inotropic and chronotropic effect and in both cases the effect is dose-dependent, slow in onset and sustained.
2. The positive inotropic response reflects the ability of endothelin-1 to increase cytosolic Ca^{2+}, an event which involves an increased influx of Ca^{2+} ions through indirectly activated calcium channels, together with the mobilization of internally stored Ca^{2+}. Mobilization of the internally stored Ca^{2+} is probably a phosphoinositol-triggered process.

3. The endothelins may contribute to, or exacerbate, the Ca^{2+} overloading which occurs during prolonged episodes of ischaemia, and during post-ischaemic reperfusion. To some extent this exacerbation may be caused by the externalization of endothelin-1 binding sites, together with the coronary constrictor effect of the polypeptide.
4. High affinity binding sites for endothelin-1 are present in the sinoatrial node, as well as in the atria and ventricles.

Chapter 15

Endothelin, Cell Proliferation and Atherosclerosis

> *"What did they live on?" said Alice, who always took a great interest in questions of eating and drinking. "They lived on treacle" said the Dormouse. "They couldn't have done that, you know" Alice gently remarked: "they'd have been ill". "So they were," said the Dormouse "very ill."*
> From: The Mad Hatters Tea Party in Alice in Wonderland: by LEWIS CAROL.

Tissue injury, macrophage infiltration, abnormal proliferation and migration of vascular smooth muscle cells, hypercholesterolaemia and the resultant accumulation of excess lipids, together with excess calcium deposition, are all intimately involved in the atherosclerotic process. The trigger for the development of the lesion is probably best accounted for in terms of the "response to injury" hypothesis originally proposed by Ross (Ross 1986). According to this hypothesis a variety of "injury promoted" growth factors – including platelet derived growth factor (PDGF), epidermal growth factor (EGF) and transforming growth factor β (TGFβ) stimulate both the migration of medial vascular smooth muscle cells into the arterial intima and their subsequent replication (Ross 1986; Schwartz et al. 1986). The lipid-loading of these cells and the subsequent formation of the fibrous "cap" are really only the terminal events (Figure 15.1).

During the past decade many different pharmacological interventions have been developed with the hope of either preventing, or at worst, slowing, the sequence of events shown in Figure 15.1. These interventions can be divided into those which are aimed at:

(i) lowering plasma cholesterol – particularly the low density plasma lipids; and
(ii) lowering blood pressure, thereby reducing the likelihood of endothelial injury, or
(iii) directly protecting the endothelial cells of the vasculature, thereby preserving their integrity, and
(iv) slowing the proliferation and migration of the smooth muscle cells. The calcium antagonists fit categories (ii), (iii) and (iv).

Atherosclerosis is widespread in the western communities. Stress, hypertension, cigarette smoking and a high cholesterol intake are undoubtedly contributory factors. But over and above this natural incidence of atherosclerosis a new problem is emerging – that of accelerated atherosclerosis in cardiac transplant patients. This problem is encountered in as many as 30–40 percent of transplant patients and remains an unsolved problem. Its cause is probably multifactorial but the possibility of the endothelins being involved can no longer be ignored. There are three reasons for reaching this conclusion.

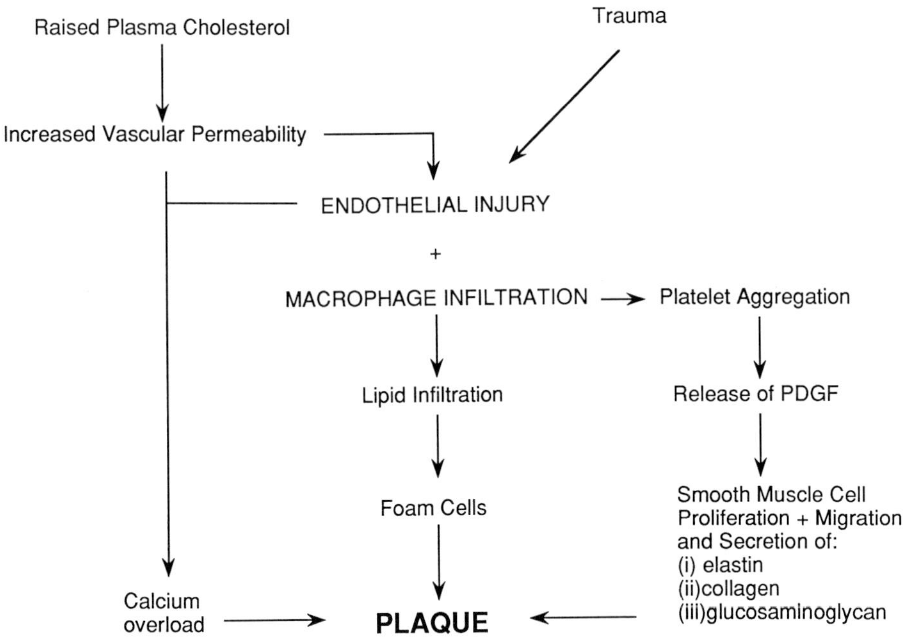

Fig. 15.1. Schematic representation of the events involved in the development of an atherosclerotic lesion

Firstly, endothelin-1 is a potent mitogen, particularly for vascular smooth muscle cells.

Secondly, as described in chapter 7, cyclosporine therapy increases the density of endothelin-1 binding sites (Nayler et al., 1989b).

Thirdly, surgical stress results in increased levels of circulating endothelin-1. It is not impossible, therefore, for endothelin-1 to be involved in the accelerated rate of atherosclerosis which is encountered after cardiac transplantation. In addition, because of its direct stimulant effect on smooth muscle cell proliferation it is equally reasonable to assume that endothelin-1 may contribute to the general phenomenon of atherosclerosis.

Endothelin-1 and the Atherosclerotic Process

Many investigators will undoubtedly argue that it is premature to involve the endothelins in the atherosclerosis saga, but such a negative approach ignores some rather compelling reasons for including them as causative or precipitating factors. Thus:

(i) because of it's ability to produce sustained vasoconstriction, even when present at low dose levels, endothelin-1 may contribute to, or even precipitate, the vessel wall injury which triggers macrophage infiltration together with smooth muscle cell proliferation and migration – processes which are fundamentally involved in the formation of atherogenic lesions.

(ii) The ability of endothelin-1 to stimulate the release of other vasopressor agents – including vasopressin, noradrenaline and adrenaline will increase the likelihood of tissue injury due to enhanced shear stress. This may be a self-perpetuating process, because shear stress stimulates endothelin mRNA (Yoshizumi et al. 1989).

(iii) Endothelin-1 has a direct mitogenic effect on vascular smooth muscle cells (Hirata et al. 1989d). In addition,

(iv) Endothelin-1 potentiates the mitogenic activity of other mitogenic agents – including the epidermal growth factor (EGF), and the transforming growth factor α.

(v) Many of the substances which are released in response to tissue injury – including thrombin, the platelet derived growth factor, transforming growth factor β, adrenaline and the interleukin-1 peptide which is released by macrophages (Yanagisawa et al. 1988a; Yoshizumi et al. 1990) all stimulate the expression of endothelin-1 mRNA. Finally,

(vi) as indicated by an increase in c-fos and c-myc mRNA levels, endothelin-1 has a direct stimulatory effect on the expression of these protooncogenes (Komura et al. 1988a, b).

The ability of endothelin-1 to act as a mitogenic agent is not restricted to its effect on vascular smooth muscle cells. The astrocytic glial cells in the brain (Supattapone et al. 1989) respond in a similar way – with marked increases in [^{3}H]thymidine incorporation accompanied by raised tissue levels of Ca^{2+}. The mesangial cells of the kidney exhibit a similar response (Simonson and Dunn 1990), as do Swiss 3T3 fibroblasts (Takuwa et al. 1989; Brown and Littlewood 1989). In each of these cases the response has been quantitated in terms of an increased uptake of [^{3}H]thymidine, preceded by an enhanced rate of phosphoinositol metabolism and a raised cytosolic Ca^{2+} . Moreover, the stimulant effect of endothelin-1 on [^{3}H]thymidine uptake is:

(i) dose-dependent, and

(ii) occurs in the same concentration range as is required to evoke a vasoconstrictor response. For rat mesangial cells, for example, the EC_{50} for the endothelin-1-induced increase in [^{3}H]thymidine uptake is around 0.9 nM (Simonson and Dunn 1990), an EC_{50} which roughly parallels that obtained for vasoconstriction (Chapter 5).

In general, therefore, there is strong background information in favour of the hypothesis that the endothelins may be one of the mediators of the atherogenic process.

The Mitogenic Effect of Endothelin-1 on Rat Vascular Smooth Muscle Cells

The abnormal proliferation of vascular smooth muscle is critically involved in the atherogenic process. Rates of proliferation can be followed by monitoring rates

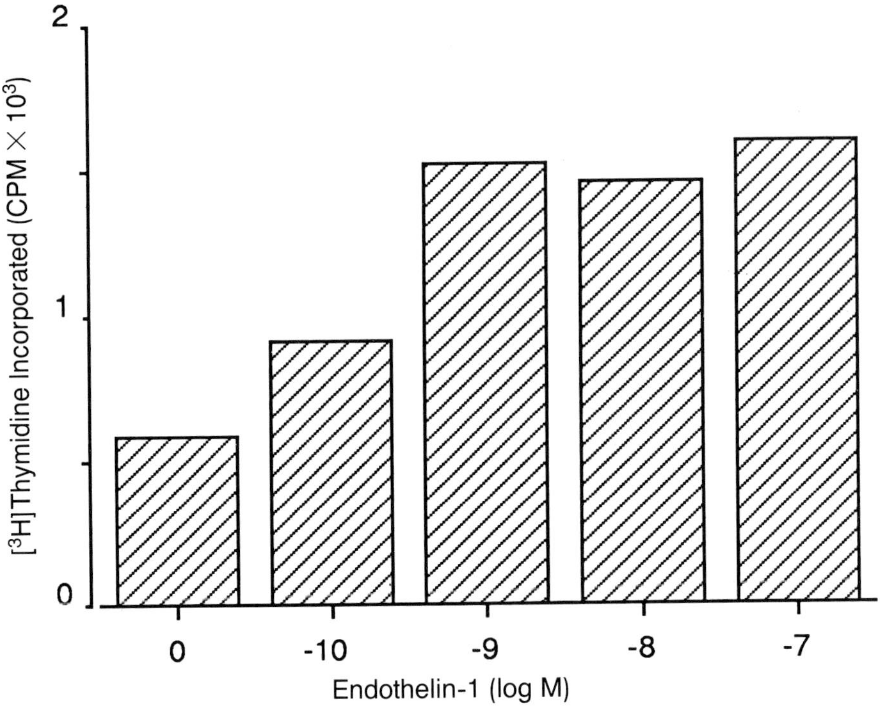

Fig. 15.2. Effect of endothelin-1 on DNA synthesis in cultured rat aortic vascular smooth muscle cells, as measured by the incorporation of [^{3}H]thymidine. (Data from Hirata et al. 1988d)

of DNA synthesis which in turn can be followed by monitoring the rates of [^{3}H]thymidine incorporation. As can be seen from Figure 15.2, exposing cultured vascular smooth muscle cells to nanomolar concentrations of endothelin-1 causes a substantial increase in [^{3}H]thymidine uptake, indicative of an accelerated rate of DNA synthesis, and hence of cell growth.

Effect of Growth Factors on the Endothelin-1 Induced Increase in [^{3}H]Thymidine Uptake

Vascular Smooth Muscle

Growth factors, including the epidermal growth factor (EGF), the transforming growth factor α (TGFα), and platelet-derived growth factor (PDGF) all stimulate DNA synthesis – and hence proliferation of vascular smooth muscle cells. The effects of EGF and TGFα are evident from the data shown in Figure 15.3, where these factors can be seen to increase [^{3}H]thymidine uptake. Of interest to the present topic, however, is the fact that two of these growth factors – TGFα and

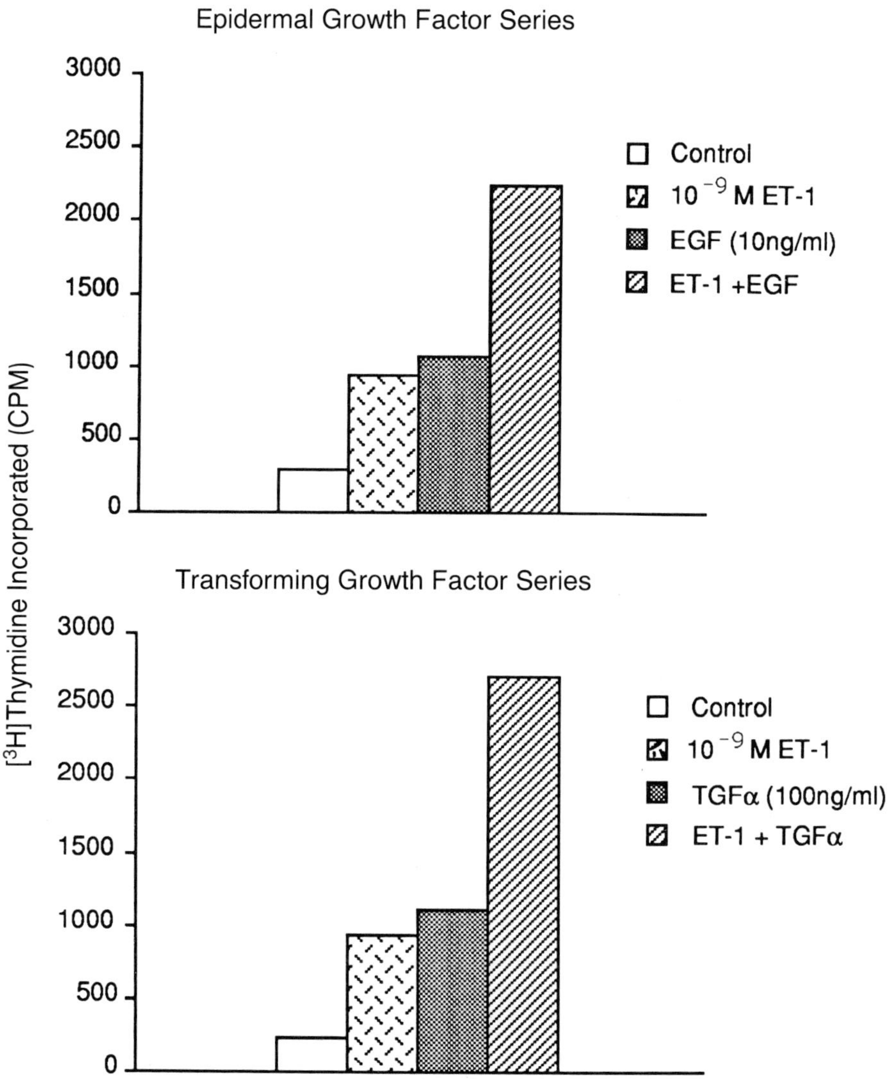

Fig. 15.3. Synergistic effect of epidermal growth factor (EGF) and transforming growth factor α (TGFα) on endothelin-1 (ET-1) induced increase in DNA synthesis as measured by the rate of [^{3}H]thymidine incorporation into cultured rat aortic vascular smooth muscle cells. (Data from Hirata et al. 1988d)

EGF exert a synergistic effect on the ability of endothelin-1 to stimulate DNA synthesis – as gauged by their joint effects on [^{3}H]thymidine uptake. Thus, even when nanomolar concentrations of endothelin-1 are present, either TGFα or EGF accelerates the rate of uptake relative to that obtained with endothelin-1 alone (Figure 15.3).

The question naturally arises as to why endothelin-1 exerts such a synergistic effect (Figure 15.3). The most likely explanation revolves around the fact that endothelin-1 raises vascular smooth muscle Ca^{2+} (Marsden et al. 1989a). A raised cytosolic Ca^{2+} is known to be an important component of the signal transduction pathway used by the growth factors (Rozengurt 1986). Hence, because endothelin-1 increases intracellular Ca^{2+}, it most probably provides conditions under which these other growth factors are able to stimulate DNA synthesis – and hence smooth muscle cell proliferation.

Other Cell Types

The ability of endothelin-1 to promote cell growth is not limited to vascular smooth muscle cells. Other examples include the Swiss ST3 cell line of fibroblasts, where endothelin-1 potentiates the ability of a variety of agents – including platelet-derived growth factor, basic fibroblast growth factor, and insulin (Kusuhara et al. 1989) to stimulate DNA synthesis. Two factors seem to be involved – the raised cytosolic Ca^{2+}, and the activation of the protein kinase C enzyme (Takuwa et al. 1989). As far as the putative endothelin-1 component of the atherosclerotic response is concerned three endothelin-induced responses may be of importance. These are its ability to:

(i) enhance Ca^{2+} influx;

(ii) mobilize intracellular Ca^{2+}; and

(iii) augment phosphoinositol metabolism, resulting in increased rates of IP turnover.

Factors Which Exacerbate the Effect of Endothelin-1 on Vascular Smooth Muscle Cell Proliferation

Factors which exacerbate the effect of endothelin-1 on smooth muscle cell proliferation can be divided into two groups. These are:

(i) factors, such as transforming growth factor, which act as synergists for the stimulant effect of endothelin-1; and

(ii) factors – including thrombin, adrenaline and the Ca^{2+} ionophore A23187 – which stimulate the production of ET-1 mRNA in vascular smooth muscle cells. Interleukin-1, a cytokine which is produced by macrophages also acts in this way (Yoshizumi et al. 1990).

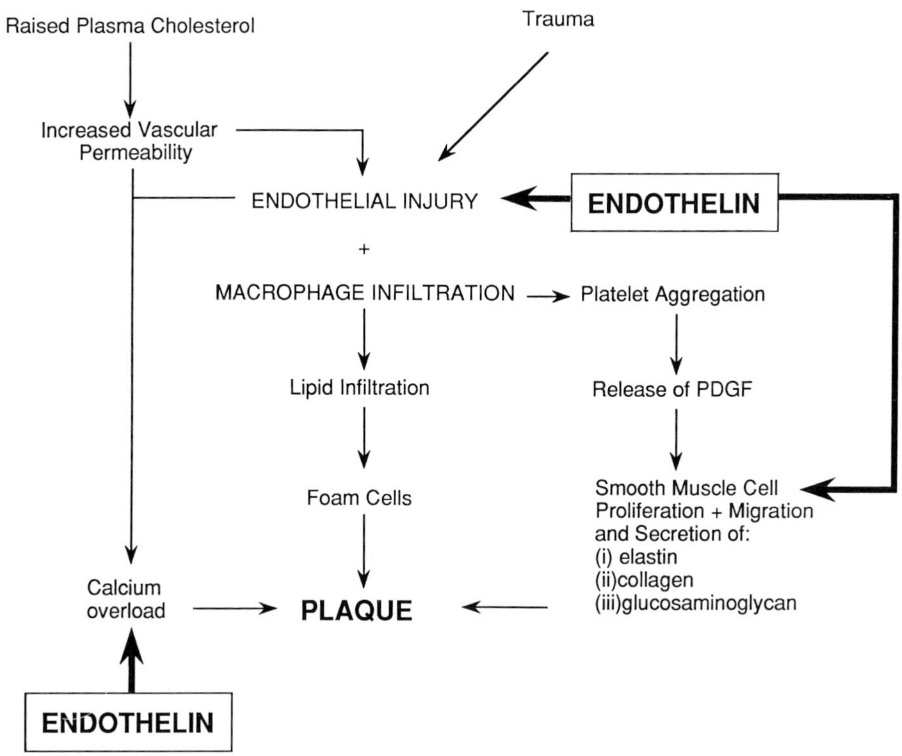

Fig. 15.4. Schematic representation of the possible involvement of endothelin in the events involved in the atherosclerotic process

The Ability of Calcium Antagonists to Attenuate Endothelin-1 Induced Cell Proliferation

Vascular Smooth Muscle

Already two reports have appeared showing that calcium antagonists are capable of attenuating the endothelin-1 induced proliferation of vascular smooth muscle cells. In 1988, Komuro and his colleagues (Komuro et al. 1988a, b) showed that nanomolar concentrations of nifedipine blocked the dose-dependent stimulant effect of endothelin-1 on prooncogene expression and DNA synthesis in rat vascular smooth muscle cells. In the same year Hirata and his colleagues (Hirata et al. 1988d) showed that nifedipine – again in nanomolar concentrations – blocked the endothelin-1 induced increase in DNA synthesis in cultured rat aortic vascular smooth muscle cells. Such an effect can be seen in Figure 15.5. These findings are of great importance, because they provide a possible biochemical explanation for the ability of the calcium antagonists to slow atherogenesis not only in laboratory studies (Henry 1985; Nayler 1988) but also in the clinical trials which are now being

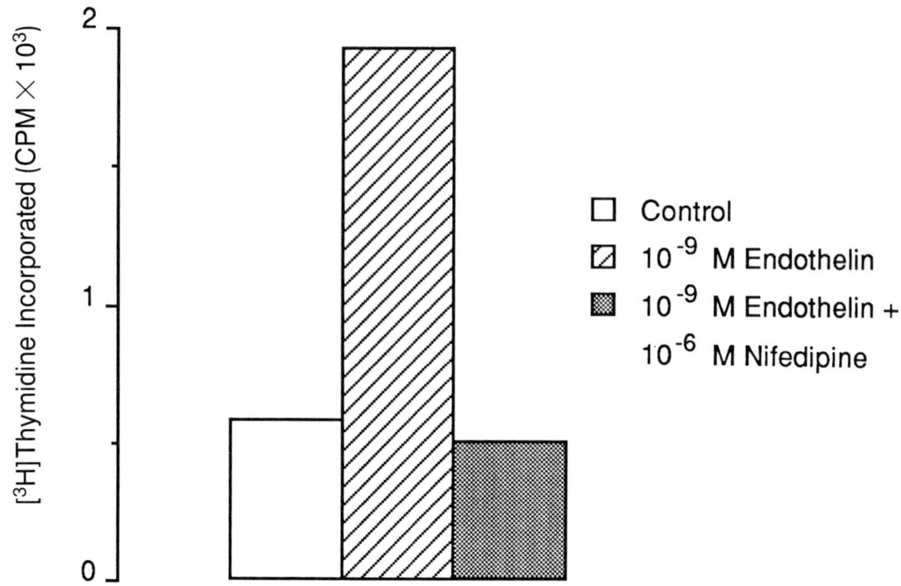

Fig. 15.5. Inhibitory effect of nifedipine (10^{-6}M) on endothelin-1 induced stimulation of DNA synthesis in cultured rat aortic vascular smooth muscle cells. DNA synthesis was monitored by measuring the rate of [^{3}H]thymidine incorporation into quiescent cells. (Data from Hirata et al. 1988d)

undertaken and which, when assessed on a three year basis, are showing a reduction in the growth of new atherogenic lesions (Lichtlen et al. 1990).

Glial Cells

The ability of calcium antagonists to antagonize the endothelin-1 induced increase in rate of cell proliferation is not peculiar to vascular smooth muscle cells, because nifedipine has a similar inhibitory effect on the endothelin-1-induced proliferation of rat cerebellar glial cells – as gauged by the rate at which they accumulate [^{3}H]thymidine (Supattapone et al. 1989).

Summary

1. The endothelins may be involved in the development of atherogenic lesions in the normal population and after cardiac transplantation.
2. The involvement of endothelin-1 (Figure 15.4) depends, in part, on
   (a) its mitogenic effect which in turn results in vascular smooth muscle cell proliferation.

(b) Endothelin-1 also exacerbates endothelial damage due to increased shear stress. This effect involves both the direct constrictor effect of the polypeptide and its ability to promote the release of other pressor agents, including vasopressin.

3. The mitogenic effect of endothelin-1 can be subdivided into
   (a) its own inherent mitogenic activity, and
   (b) its ability to potentiate the mitogenic activity of other agents – including the epidermal growth factor and transforming growth factor α.
4. Substances released in response to tissue injury (platelet derived growth factor, transforming growth factor β) as well as interleukin-1 derived from macrophages stimulate the synthesis of endothelin-1.
5. Nanomolar concentrations of the calcium antagonist nifedipine block the dose-dependent stimulatory effect of endothelin-1 on vascular smooth muscle cell proliferation. This effect probably contributes to the antiatherosclerotic effect of nifedipine reported in a recent clinical trial (Lichtlen et al. 1990).

Chapter 16

The Constrictor Effect of the Endothelins on Bronchiolar and Tracheal Smooth Muscle

"And in this harsh world, draw thy breath - in pain".
W. SHAKESPEARE: in Hamlet

Shakespeare might have been describing the consequences of inhaling endothelin-1, because irrespective of whether the polypeptide is administered by aerosol (Lagente et al. 1989) or it is applied topically (Uchida et al. 1988; Hay 1989; Turner et al. 1989a, b) bronchiolar and tracheal smooth muscle respond by developing a sustained contraction. The response has all the hallmarks of an endothelin-induced contraction because:

(i) it is dose-dependent (Figure 16.1);
(ii) it is often preceded by a transitory relaxation;
(iii) it is sustained, and is accompanied by a raised cytosolic Ca^{2+} (Sakata et al. 1989);
(iv) it is dependent, in part, on the presence of extracellular Ca^{2+},
(v) it is attenuated, but not completely abolished, by calcium antagonists (Table 16.1 and Lee et al. 1990), and

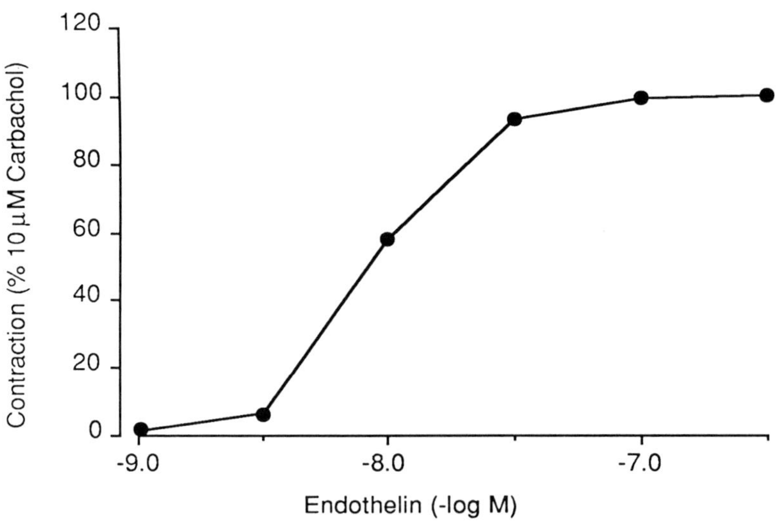

Fig. 16.1. Dose-response curve for the constrictor effect of endothelin-1 on isolated guinea pig trachea. Results are expressed as a percentage of the contractile response to 10 μM carbachol. Note that the threshold for the constrictor effect of endothelin-1 is around 50 nM

Table 16.1. Effect of calcium antagonists on the constrictor effect of endothelin-1 on tracheal and bronchiolar smooth muscle

Preparation	Calcium Antagonist	Change in Endothelin-1 response	Reference
Sheep trachea	Gallopamil	Partial inhibition	Ryan et al. 1989
Guinea pig trachea	Nicardipine	Partial inhibition	Uchida et al. 1988
Guinea pig bronchi	Nifedipine	Partial inhibition	Maggi et al. 1989a
Dog trachea	Verapamil	Partial inhibition	Sakata et al. 1989
Ferret trachea	Verapamil	Partial inhibition	Lee et al. 1990

Note that the inhibitory effect of the calcium antagonists is a class effect and not dependent upon their chemical structure, nicardipine and nifedipine being dihydropyridines and gallopamil and verapamil, phenylalkylamines

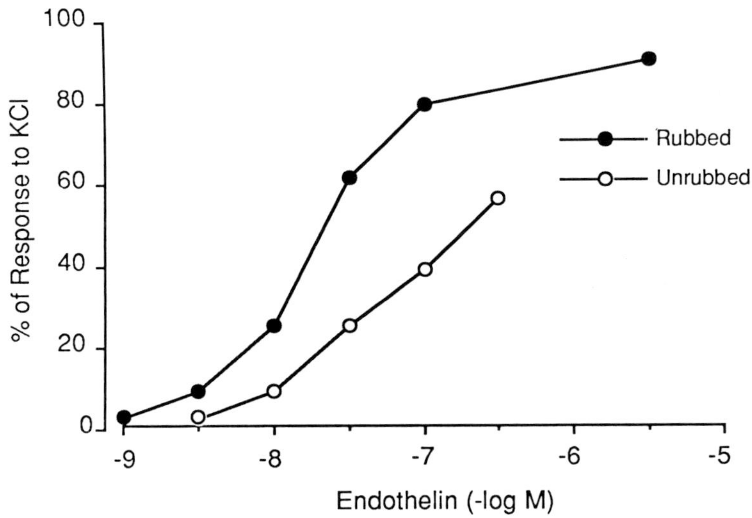

Fig. 16.2. Concentration response curve for the constrictor effect of endothelin-1 on isolated guinea pig bronchi. Pairs of bronchi were used, one serving as a control, the other being rubbed to remove the endothelium. Note that removal of the endothelium moves the dose-response curve for the bronchoconstrictor effect of endothelin-1 to the left. (Data from Maggi et al. 1989a)

(vi) in both bronchiolar and tracheal muscle, concentrations of endothelin-1 which evoke a constrictor response partially depolarize the cell membrane (Lee et al. 1990).

In addition,

(vii) removing the endothelium enhances the constrictor response (Figure 16.2).

There are several reasons why endothelial denudation might enhance the bronchoconstrictor effect of the endothelins. Firstly, removal of the endothelium automati-

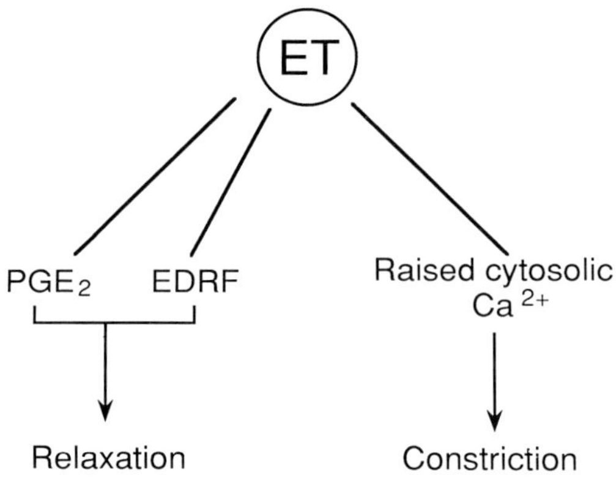

Fig. 16.3. Schematic representation of the effect of the endothelins on bronchiolar and tracheal smooth muscle. Note that the end result represents the balance between the direct constrictor effect of the endothelins and their ability to stimulate the release of endothelial-derived relaxing agents

cally removes the degradative peptidases which are contained within these cells. Secondly, the endothelins provoke the release of relaxant factors, including some prostanoids and EDRF, from the endothelium (Maggi et al. 1990). Hence, when an endothelin is added to a bronchiolar smooth muscle preparation with an *intact* endothelium, the resultant constrictor response is counterbalanced, to some extent, by the relaxant effect of the released endothelial-derived relaxing agents (Figure 16.3). This effect seems to be more marked for endothelin-3 (Maggi et al. 1990) than endothelin-1, so much so that when denuded preparations are used the bronchoconstrictor potency of endothelin-3 sometimes exceeds that of endothelin-1.

The ability of the endothelins to stimulate the release of endothelial-derived relaxing factors from respiratory epithelial cells raises some interesting possibilities. For example, are there two types of endothelin receptors in the respiratory system? May be one type is located on the smooth muscle cells where it could be involved in the cascade of events which results in the raised cytosolic Ca^{2+} involved in the constrictor response. The second type of receptor, if it exists at all, might be located on the epithelial cells and be concerned with the events involved in the endothelin-induced release of the relaxing agents.

The Relative Sensitivity of Bronchiolar and Tracheal Smooth Muscle to the Constrictor Effect of the Endothelins

Tracheal and bronchiolar smooth muscle are approximately 50–100 times *less* sensitive to endothelin-1 than aortic smooth muscle (Turner et al. 1989b). This difference in the relative sensitivity of tracheal and aortic smooth muscle is apparent in the tension data shown in Figure 16.4. Despite these relative differences in sensitivity, however, endothelin-1 is still a potent constrictor of tracheal and bronchiolar muscle, the maximum response being almost equal to that of histamine and nearly double that of leukotreine D_4 (Uchida et al. 1988).

In endothelial-denuded preparations the constrictor effect of endothelin-3 $>$ endothelin-1 (Maggi et al. 1990).

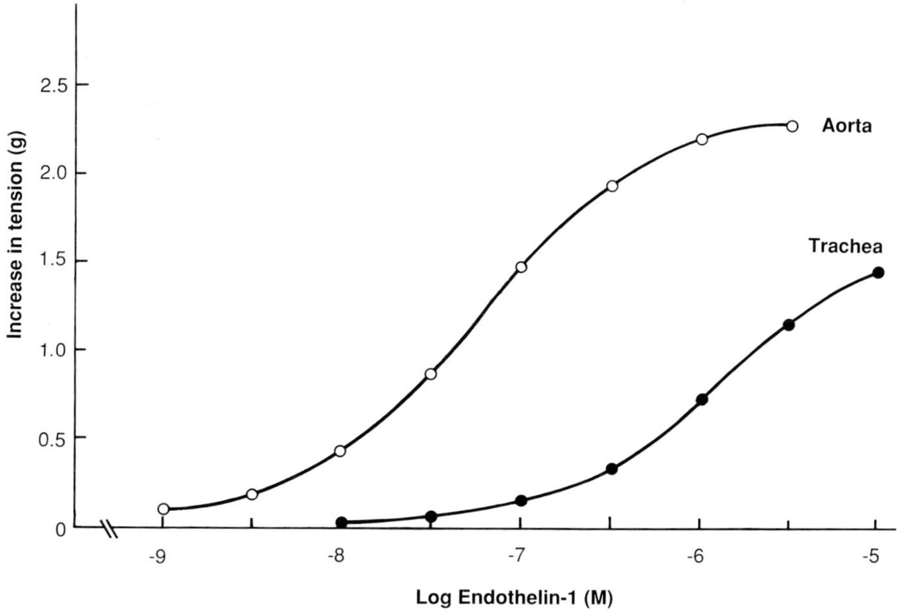

Fig. 16.4. Dose response curves for the constrictor effect of endothelin-1 on guinea pig tracheal and aortic smooth muscle preparations. Each point is the mean of six experiments. The perfusion fluid was Krebs-Henseleit buffer. Cumulative doses of endothelin-1 were used

Table 16.2. Bronchiolar epithelial cells which produce endothelin-like polypeptides

Tissue			Endothelin-like immunoreactivity
Epithelial cell type	(i)	serous	intense
	(ii)	mucous	intense
	(iii)	ciliated	no response/faint
	(iv)	basal	no response/faint
Small Bronchioles	(i)	Clara	intense
	(ii)	ciliated	faint

Data from Rozengurt et al. 1990

The Local Production of Endothelin-1 and Endothelin-3

Respiratory epithelial cells synthesize both endothelin-1 and endothelin-3 (Black et al. 1989; Rozengurt et al. 1990). As can be seen from Table 16.2, the sites of production are concentrated in the mucous, serous and Clara cells, rather than the basal or ciliated epithelial cells.

Mode of Action

The constrictor effect of the endothelins on respiratory smooth muscle probably involves at least two mechanisms:
(a) indirect activation of the calcium channels by membrane depolarization; and
(b) direct activation of these channels by way of a calcium antagonist-insensitive binding site.

In either case the end result is a raised cytosolic Ca^{2+} which presumably triggers the contractile response. However, since the calcium antagonists only attenuate, and do not abolish the contractile response, it is reasonable to assume that other factors (in addition to an enhanced Ca^{2+}-channel dependent influx of Ca^{2+} are involved. There are several possibilities, including
(i) enhanced phosphoinositol metabolism and the subsequent release of internally stored Ca^{2+}, and
(ii) enhanced transmitter release from sympathetic nerve terminals.

The Role of the Endothelins in Respiratory Disorders

It is too early to speculate as to whether the endothelins might contribute to conditions such as *status asthmaticus*, although the idea is attractive. An abnormal constrictor response could come about in at least three endothelin-dependent ways:
(i) there may be an excess secretion of the polypeptide(s) from the epithelial cells;
(ii) the epithelial cells may fail to release their endogenous relaxing agents so that the unopposed constrictor effect of the endothelins becomes the dominant response, and
(iii) either the density or the affinity of the receptors may change.

The presence of excess endothelin-1 in bronchial lavages obtained from patients in *status asthmaticus* (Nomura et al. 1989, Chapter 10) adds some weight to the first of these possibilities.

Irrespective of whether there are one or two populations of endothelin-specific binding sites in bronchiolar and tracheal smooth muscle as has been suggested, the ability of the adjacent epithelial cells to synthesize and secrete endothelin-1 and endothelin-3 means that the polypeptides are acting as local hormones and under conditions in which relatively high concentrations of the polypeptides might be available for receptor activation.

Summary

1. Endothelin-1 and endothelin-3 are potent constrictors of bronchiolar and tracheal smooth muscle.

2. The constrictor effect of endothelin-3 on denuded respiratory smooth muscle > endothelin-1.
3. Specific high affinity binding sites for endothelin-1 (and -3) are present in bronchiolar and tracheal smooth muscle.
4. Bronchiolar and tracheal epithelial cells synthesize endothelin-1 and endothelin-3.
5. The constrictor effect of endothelin-1 on bronchiolar and tracheal smooth muscle is attenuated (but not abolished) by calcium antagonists. It represents a balance between the direct constrictor effect of the endothelin, and the relaxant effect caused by the endothelin-induced release of relaxing agents from the endothelium.
6. The endothelins could contribute to *status asthmaticus*.

Chapter 17

Endothelin and Hypertension

"He has, indeed, done it very well; but it is a foolish thing well done."
SAMUEL JOHNSON: in the London Chronicle in his reply to Goldsmith's apology for beating Evans the bookseller. 1773

Although Johnson was referring to Goldsmith's apology for what must have been some unseemly behaviour, his words are equally applicable to the endothelins – because if:

(i) they are secreted in excess (particularly endothelin-1);
(ii) removed slowly;
(iii) their receptor density increases or their affinity changes; or
(iv) the compensatory dilator mechanisms which include the release of either
  (a) ANP (Fukuda et al. 1988),
  (b) EDRF and prostacyclin (de Nucci et al. 1988)

  fail, or diminish, then systemic vascular resistance may increase, because endothelin (endothelin-1) has a potent and sustained vasoconstrictor effect (Table 17.1).

For example, when infused into the forearm of man at the rate of 5 pmol per minute, forearm blood declines by almost sixty percent (Clarke et al. 1989). In marked contrast to the vasoconstrictor effect of many other vasoactive peptides, including angiotensin II, the effect of endothelin-1 is slow in onset and, as might have been expected from the earlier sections of this book, is sustained unless calcium antagonists are administered (Clarke et al. 1989).

Because of its slow onset of action and sustained effect, it is improbable that endothelin-1 contributes to any of the sudden changes in blood pressure which

Table 17.1. Evidence of the vasoconstrictor effect of endothelin-1

Animal	Reference
Conscious rats	Ouchi et al. 1989
Anaesthetized rats	Han et al. 1989
Anaesthetized rats	Hirata et al. 1989c
Anaesthetized dogs	Miller et al. 1989
DOCA-treated rats	de Carvalho et al. 1990
Spontaneously hypertensive rats	Miyauchi et al. 1989
Anaesthetized cats	Minkes et al. 1989
Awake rats	Madeddu et al. 1989

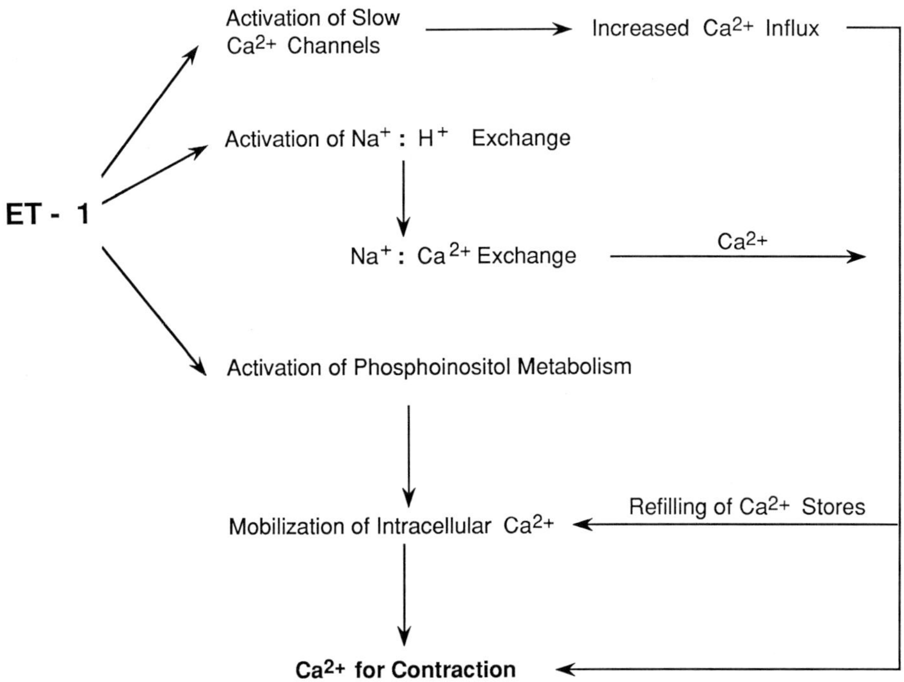

Fig. 17.1. Schematic representation of the mechanisms whereby endothelin-1 might increase the amount of Ca^{2+} available for excitation-contraction coupling in smooth muscle cells

occur in normal daily life – as in response to exercise, for example. In any case the plasma levels of endothelin-1 do not change in response to exercise (Hartter et al. 1989). Even if they did the increase would have to be substantial because whereas the plasma levels in healthy adults usually ranges between 1 and 2 pg per ml (Chapters 9 and 10) the threshold for endothelin's vasoconstrictor effect is higher than this (Chapter 5).

A vasoconstrictor effect of endothelin-1 based on a potentiation of noradrenergic transmission cannot be entirely excluded, because, although endothelin-1 either acts in the opposite manner or fails to alter stimulation-induced noradrenaline release from rat atria (Reid et al. 1989) it enhances adrenergic vasoconstriction (Tabuchi et al. 1989). However, the main event underlying the vasoconstrictor response is undoubtedly a raised cytosolic Ca^{2+} (Figure 17.1).

Plasma Endothelin-1 Levels in Hypertensives

Although the plasma levels in hypertensive patients are below the threshold needed to evoke a vasoconstrictor response, they are significantly higher than the levels found in normotensives, the increases being of the order of one hundred percent (Saito et al. 1990). However, in age-matched groups of patients the increase in plasma level correlates neither with systolic nor diastolic blood pressure. Probably, therefore, Saito and his colleagues are correct in assuming that the "elevated plasma level of endothelin-1-like immunoreactivity in patients with essential hypertension is related to vascular complications, including atherosclerosis" (Saito et al. 1990). It is possible, however, that these higher than normal levels of plasma endothelin-1 contribute to, or even explain why, plasma ANP levels rise under these conditions, since endothelin-1 promotes ANP release (Fukuda et al. 1988). In any case, hypertension *per se* has an interesting effect on the ability of atria to release ANP, the release being significantly potentiated in spontaneously hypertensive rats (Winquist et al. 1989a). Maybe this is an adaptive mechanism.

Endothelin-1 Binding Site Density in Spontaneously Hypertensive Rats

Large capacitance and renal arteries from hypertensive rats are hypersensitive to the constrictor effect of endothelin-1 (Clozel 1989; Tomobe et al. 1988). Moreover, the difference progresses with age (Miyauchi et al. 1989). There are a number of factors which could contribute to this hypersensitivity, including
(i) an increase in receptor binding site density, or affinity;
(ii) a change in the post-receptor events; and
(iii) diminished localized release of the vasodilators EDRF and prostacyclin.

The first of these possibilities – an *increase* in receptor number and/or affinity can now be discounted *as far as the vasculature* is concerned, because the number of receptors actually falls! Aortic smooth muscles from normotensive Wistar Kyoto rats, for example, contain approximately three thousand receptor sites per cell, whereas in hypertensive rats the level falls to around thirteen hundred whilst the affinity is unchanged (Clozel 1989).

As yet little information is available concerning the post-receptor events with respect to the increased sensitivity to endothelin-1, but there is indirect evidence of an adaptive mechanism coming into play, because the initial and transient depressor response which follows the intravenous administration of endothelin-1, and which is probably due to the localized release of EDRF and prostacyclin, together with ANP release from the cardiac atria, is accentuated (Winquist et al. 1989b).

Is a Central Effect of the Endothelins Involved in Hypertension?

There are two reasons for believing that this may be the case.
(i) When injected into the lateral cerebral ventricle picomolar concentrations of endothelin-1 cause a dose-dependent *increase* in mean arterial blood pressure which is due, in part, to a concommitant increase in plasma noradrenaline and adrenaline (Ouchi et al. 1989).
(ii) Membranes prepared from homogenized brains of spontaneously hypertensive rats exhibit an *increase* in endothelin-1 binding site density (Table 17.2, and Gu

Table 17.2. Effect of hypertension on endothelin-1 binding site density in cardiac and brain membranes from normotensive and spontaneously hypertensive rats

	Normotensives	Hypertensives
	B_{max} (fmol/mg protein)	
Membrane Source		
Atria	115	37
Brain	130	167

(Data from Gu et al. 1990)
The normotensive rats were of the Wistar Kyoto and the hypertensives of the Spontaneous Hypertensive strain. The hypertensive-induced increase in brain receptor density was accompanied by a small but significant *increase* in affinity whereas the opposite applied to the atrial membranes

et al. 1990) without marked changes in affinity. This is probably *not* a non-specific effect of hypertension because at the same time (Table 17.2) the atrial binding site density decreases.

Is There a Vasodilator-Vasoconstrictor Imbalance?

This is certainly possible because although the acute intravenous injection of endothelin-1 can evoke an enhanced vasodilator response in hypertensives, prolonged hypertension dampens the transient dilator response, particularly in large arteries (Luscher 1989). One interpretation of these findings is that prolonged hypertension modifies the endothelium, decreasing its capacity to respond to or secrete EDRF, and possibly prostacyclin. Under those conditions the constrictor effect of endothelin would be the dominant response. In many ways, therefore, we are back to the old YIN-YAN hypothesis.

Is the Vasoconstrictor Effect of Endothelin due to it's Effect on Large Capacitance or Small Resistance Vessels?

Irrespective of whether the enhanced vasoconstrictor effect of endothelin which has now been described for several models of hypertension (de Carvalho et al. 1990; Miyauchi et al. 1989) includes a local or central effect of the polypeptide, or both, it is important to establish whether it is the large capacitance or small resistance vessels which are primarily affected. Recent data published by de Carvalho et al (de Carvalho et al. 1990) makes a strong argument for favouring the microvessels as being the major target sites.

Mode of Action

If the endothelins do contribute to the hypertensive state, as it occurs in man, then it is necessary to explore its mode of action at the subcellular level. The following observations provide likely clues:

(i) endothelin-1 causes a sustained increase in intracellular Ca^{2+} in vascular smooth muscle (Meyer-Lehnert et al. 1989);

(ii) drugs which inhibit the Na^+:H^+ exchanger attenuate the vasoconstrictor response (Richards et al. 1989).

(iii) The response is attenuated by calcium antagonists, including nifedipine (Madeddu 1989). Based on these observations at least two routes of Ca^{2+} entry are probably involved:

  (a) excess entry through the slow Ca^{2+} channels,

  (b) entry by way of the Na^+:H^+ exchanger, secondary to the activation of the Na^+:H^+ exchanger.

Added to this is the additional Ca^{2+} which is mobilized from the intracellular stores (Figure 17.1).

Agents Which Overcome the Vasoconstrictor Effect of Endothelin-1

These agents include nitroglycerine (Miller et al. 1989), nifedipine (Madeddu et al. 1989), and nitroprusside (Ohman et al. 1989) but not saralasin.

Endothelin-1, Hypertension and the Renal Vasculature

The renal vasculature of hypertensive rats appears to be particularly sensitive to the constrictor effect of endothelin-1 (Madeddu 1989). This raises the interesting possibility of endothelin being involved not only in blood pressure, but also in body fluid regulation and in the excessive renal vascular resistance which accompanies experimentally and naturally occurring hypertension (Lowenstein et al. 1970).

Summary

1. There is increasing evidence in favour of the hypothesis that the endothelins are involved in the aetiology of hypertension.
2. The microvessels, rather than the capacitance vessels, are the probable target sites and display hypersensitivity.
3. Since the vasoconstriction is calcium antagonist-sensitive some component of the response probably involves Ca^{2+} entry through the voltage-sensitive Ca^{2+} channels.
4. It is probable that part of the vasoconstrictor effect of the endothelins involves a central effect.

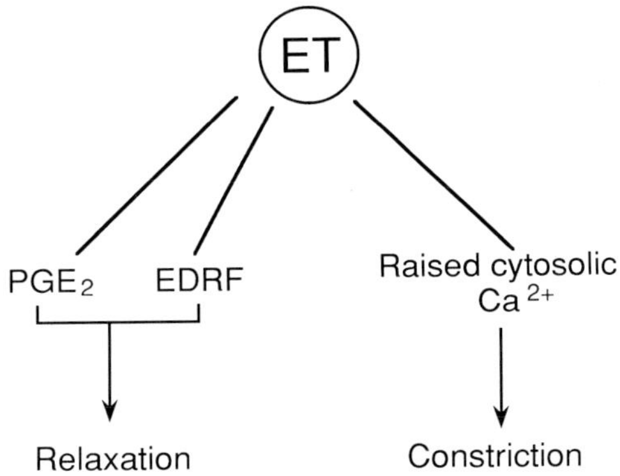

Fig. 17.2. Schematic representation of the factors which are normally involved in maintaining smooth muscle tone at the cellular level. Adding endothelin-1 to such a system will tilt the balance in favour of constriction, and hence a rise in vascular resistance

5. Although the endothelin-1 receptor density in the capacitance vessels of spontaneously hypertensive rats is reduced (relative to normotensive) the overall density in the brain increases, without change in affinity.
6. The enhanced vasoconstrictor response to endothelin-1 seen in hypertensives progresses with age.
7. The raised plasma levels of endothelin found in hypertensives is inadequate to account for the hypertensive state, the levels remaining below the level needed for contraction. However the raised plasma levels probably reflect a much larger increase in the endogenous levels of the polypeptide.
8. Possibly the endothelins (endothelin-1) contribute to the hypertensive state by adding another constrictor component to a system which is already hyperactive, possibly because of a regionalized imbalance (Figure 17.2) between the localized release of dilator agents (prostacyclins and EDRF) and metabolic disturbances which raise cytosolic Ca^{2+}.

Chapter 18

The Endothelins and Platelet Aggregation

> *"I hope you are not leading a double life, pretending to be wicked and being really good all the time".*
> The Importance of Being Earnest – OSCAR WILDE

Oscar Wilde might have been referring to the endothelins, since on the one hand they have the potential for causing or exacerbating systemic hypertension, renal artery stenosis, coronary artery occlusion, cerebral vasospasm, and atherosclerosis whilst on the other they inhibit platelet aggregation (Thiemermann et al. 1990) and can be fibrinolytic (Korbut et al. 1989). Except under conditions of severe trauma, when excessive blood loss might be a problem, the ability of these polypeptides to slow platelet aggregation and to activate the fibrinolytic system, must surely be regarded as desirable.

The Inhibitory Effect of Endothelin-1 on Platelet Aggregation

Platelets, like red blood cells, lack specific binding sites for endothelin-1 (Takayanagi et al. 1990). It follows, therefore, that any anti-aggregatory effect which this polypeptide might exert must be a secondary phenomenon. This conclusion is supported by the fact that when added to platelet-rich plasma under *in vitro* conditions, endothelin-1 has no effect on platelet aggregation (de Nucci et al., 1988). However, when injected either intra-arterially or intravenously, before the platelet rich plasma is withdrawn and assayed for platelet aggregability (Thiemermann et al. 1989), or platelet aggregability is assessed under *in vivo* conditions by monitoring the accumulation of radioactively labelled platelets in a particular vascular bed (Thiemermann et al. 1990) endothelin-1 inhibits platelet aggregation. The effect is evident within a few minutes, and is sustained (Figure 18.1). Moreover it is only attenuated, and not abolished, by the prior injection of the cyclooxygenase inhibitor indomethacin (Thiermermann et al. 1988).

The most likely explanation for these findings is that:

(i) the endothelin-1 induced inhibition of platelet aggregation is a secondary phenomenon, involving the release of more than one anti-aggregating agent;

(ii) since the response is only attenuated, and not abolished by indomethacin, the effect can only be accounted for in part, in terms of prostacyclin release (Moncada et al. 1976; Thiermermann et al. 1989).

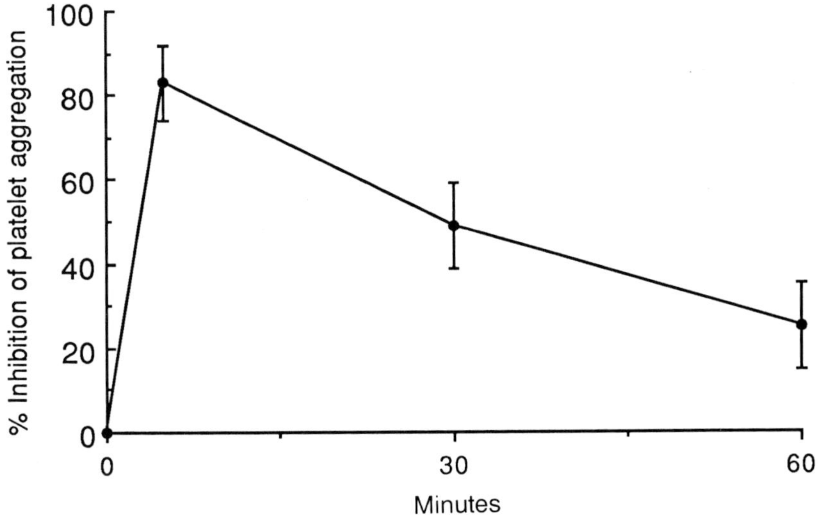

Fig. 18.1. Percentage inhibition of platelet aggregation caused by the intra arterial injection of 1 nmol/kg endothelin into anaesthetized rabbits. After injecting the endothelin, platelet rich plasma was withdrawn and assayed for platelet aggregability, using 2.56 μM adenosine diphosphate to trigger aggregation. (Data from Thiermermann et al. 1988)

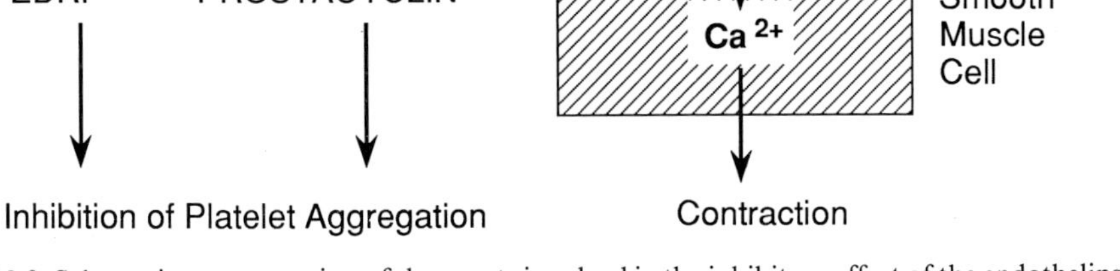

Fig. 18.2. Schematic representation of the events involved in the inhibitory effect of the endothelins on platelet aggregation

Since endothelin-1 also releases EDRF (de Nucci et al. 1988; Warner et al., 1988, 1989; Rae et al. 1989) and because EDRF inhibits platelet aggregation (Azuma et al. 1986; Furlong et al. 1987) and platelet adhesion (Radomski et al. 1987; Sneddon and Vane 1988), and the inhibitory effect of EDRF on platelet aggregation is insensitive to indomethacin it seems likely that EDRF, as well as prostacyclin contributes to the inhibitory effect of endothelin-1 on platelet aggregation. Possibly the situation resembles that shown in Figure 18.2.

The involvement of prostacyclin and EDRF in the anti-platelet aggregatory effect of endothelin-1 raises several questions:

(i) is the inhibitory effect peculiar to endothelin-1?

(ii) does the vasoconstrictor effect of endothelin-1 trigger the release of the prostacyclin and EDRF?

(iii) how do the endothelins promote the release of these antiaggregatory compounds? Is it a receptor mediated event and if so, are there a separate subpopulation of receptors involved?

The first two of these questions can be answered in the negative, because endothelin-3, which has only one twentieth of the vasoconstrictor effect of endothelin-1 (Yanagisawa et al. 1988a, b) but which is equipotent in releasing EDRF, also inhibits platelet aggregation (Lidbury et al. 1989). Moreover the cyclooxygenase inhibitors, whilst inhibiting prostacyclin production, attenuate the effect of endothelin-3 on platelet aggregation without modifying its effect on perfusion pressure. The third question can probably be answered in the affirmative, because whereas endothelin-3 $<$ endothelin-1 in terms of its vasoconstrictor potency, it is equipotent in terms of its EDRF releasing activity. This implies that different receptors must be involved in the two processes – one subset for contraction, the other for EDRF, and presumably prostacyclin, rclcase.

The Fibrinolytic Activity of Endothelin-3

As well as slowing platelet aggregation, endothelin-3 is also fibrinolytic (Korbut et al. 1989). This effect is long lasting and seems to involve the ability of endothelin-3 :

(i) to stimulate prostacyclic production;

(ii) to release EDRF, and

(iii) to release plasminogen activator from the endothelium.

Summary

1. Endothelin-1 and endothelin-3 inhibit platelet aggregation. Endothelin-3 also inhibits fibrinolytic activity.
2. These effects are secondary and probably involve the release of
   (a) prostacyclin;
   (b) EDRF, and
   (c) plasminogen activator.
3. The endothelin-1 and endothelin-3 induced release of either prostacyclin, EDRF or plasminogen activating factor is *not* dependent upon an endothelin-induced increase in perfusion pressure.

Chapter 19

Do the Endothelins have a Future?

"Should we force science down the throats of those that have no taste for it? Is it our duty to drag them kicking and screaming into the twenty first century? I am afraid it is."
SIR GEORGE PORTER, 1986. Nobel Laureate

There can be little doubt now that investigators will continue to probe the mysteries of the endothelins until their properties are fully elucidated, their mode of action thoroughly exposed, their receptors cloned, and their patho- and physiological significance determined. Explorations will not stop here, however, because if these polypeptides are involved in certain pathologies – including cerebral vasospasm, renal failure, hypertension, myocardial infarction, atherosclerosis, mental depression and possibly neurosecretory abnormalities then it will be important to establish whether a specific receptor antagonist can be developed – or should the effort be directed towards searching for substances which will inhibit the activity of the converting enzyme which is responsible for releasing the biologically active "mature" endothelin from its precursor state. Experience with inhibitors of the angiotensin converting enzyme has shown that such an approach is both practical, and profitable. On the other hand experience in the beta adrenoceptor and calcium antagonist field has shown that specific receptor antagonists have wide therapeutic applications.

As far as the endothelins are concerned many factors will have to be borne in mind when searching for suitable antagonists, including the existence of receptor sub-types and the rapidity with which the receptors "up" and "down" regulate.

In the meantime, until specific antagonists, or converting enzyme inhibitors, are developed it may be pertinent to remember that although these polypeptides do not interact directly with the voltage sensitive Ca^{2+} channels, an indirectly enhanced influx of Ca^{2+} through these channels plays a primary role in their mode of action. Because of this the currently available calcium antagonists provide at least one therapeutically acceptable way of regulating the physiological, and possibly pathological consequences of endothelin secretion. Is this why the calcium antagonists slow atherogenesis, and does it contribute to their ability to lower blood pressure in hypertensive but not normotensive patients? These and many other questions will probably be answered in the next few years. In the meantime it is difficult to realise that it is only two years since the structure of the endothelin prototype – endothelin-1 – was determined.

The next two years may prove to be even more exciting, as the physiological and pharmacological properties of these polypeptides become more clearly identified, and their interaction with other regulatory systems is identified.

References

Ambar I, Kloog Y, Schvartz I, Hazum E, Sokolovsky M (1989a) Competitive interaction between endothelin and sarafotoxin: binding and phosphoinositides hydrolysis in rat atria and brain. Biochem Biophys Res Commun 158:195–201

Ambar I, Kloog Y, Sokolovsky M (1989b) Solubilization of endothelin/sarafotoxin receptors in an active binding form. Eur J Pharmacol 170:119–120

Ando K, Hirata Y, Shichiri M, Emori T, Marumo F (1989) Presence of immunoreactive endothelin in human plasma. FEBS Lett 245:164–166

Anggärd E, Galton S, Rae G, Thomas R, McLoughlin L, de Nucci G, Vane JR (1989) The fate of radioiodinated endothelin-1 and endothelin-3 in the rat. J Cardiovasc Pharmacol 13 (Suppl 5) S46–S49

Asano T, Ikegaki I, Suzuki Y, Satoh S-I, Shibuya M (1989) Endothelin and the production of cerebral vasospasm in dogs, Biochem Biophys Res Commun 159:1345–1351

Auguet M, Delaflotte S, Chabrier P-E, Pirotzky E, Clostre F, Braquet P (1988) Endothelin and Ca^{2+} agonist Bay K8644: different vasoconstrictive properties. Biochem Biophys Res Commun 156:186–192

Azuma H, Ishikawa M, Sekikaki S (1986) Endothelium-dependent inhibition of platelet aggregation. Brit J Pharmacol 88:411–415

Badr KF, Munger KA, Sugiura M, Snajdar RM, Schwartzberg M, Inagami T (1989) High and low affinity binding sites for endothelin on cultured rat glomerular mesangial cells. Biochem Biophys Res Commun 161:776–781

Baydoun AR, Peers SH, Cirino G, Woodward B (1989) Effects of endothelin-1 on the rat isolated heart. J Cardiovasc Pharmacol 13 (Suppl 5):S193–S196

Bdolah A, Wollberg Z, Fleminger G, Kochva E (1989) SRTX-d, a new natural peptide of the endothelin/sarafotoxin family. FEBS Lett 256:1–3

Berbinschi A, Ketelslegers JM (1989) Endothelin in urine. Lancet ii:46

Berridge MJ, Irvine RF (1989) Inositol phosphates and cell signalling. Nature (Lond) 341:197–204

Bkaily G, Sperelakis N, Renaud JF, Payet MD (1985) Apamin, a highly specific Ca^{2+} blocking agent in heart muscle. Am J Physiol 248:H961–H965

Black PN, Ghatei MA, Takahashi K, Bretherton-Watt D, Krausz T, Dollery CT, Bloom SR (1989) Formation of endothelin by cultured airway epithelial cells. FEBS Lett 255:129–132

Bloch KD, Eddy RL, Shows TB, Quertermous T (1989a) cDNA cloning and chromosomal assignment of the gene encoding endothelin-3. J Biol Chem 264:18156–18161

Bloch KD, Friedrich SP, Lee ME, Eddy RL, Shows TB, Quertermous T (1989b) Structural organization and chromosomal assignment of the gene encoding endothelin. J Biol Chem 264:10851–10857

Borges R, Carter DV, von Grafenstein H, Halliday J, Knight DE (1989) Ionic requirements of the endothelin response in aorta and portal vein. Circ Res 65:265–271

Boulanger C, Luscher TF (1990) Release of endothelin from the porcine aorta. Inhibition by endothelium-derived nitric oxide. J Clin Invest 85:587–590

Bousso-Mittler D, Kloog Y, Wollberg Z, Bdolah A, Kochva E, Sokolovsky M (1989) Functional endothelin/sarafotoxin receptors in the rat uterus. Biochem Biophys Res Commun 162:952–957

Brain SD, Tippins JR, Williams TJ, (1988) Endothelin induces potent microvascular constriction. Brit J Pharmacol 95:1005–1007
Brain SD (1989) The direct observation of arteriolar constriction induced by endothelin in vivo. Eur J Pharmacol 160:401–403
Brown KD, Littlewood CJ (1989) Endothelin stimulates DNA synthesis in Swiss 3T3 cells. Biochem J 263:977–980
Cavero PG, Miller WL, Aarhus LL, Heublein DM, Burnett JC (1989) Acute congestive heart failure results in a selective attenuation of endothelial-mediated vasoconstriction in the kidney. Am J Hypertens 2:Abst 38A
Cernacek P, Stewart DJ (1989) Immunoreactive endothelin in human plasma: marked elevations in patients in cardiogenic shock. Biochem Biophys Res Commun 161:562–567
Chai SY, Sexton PM, Allen AM, Figor R, and Mendelsohn FAO (1986) In vitro autoradiographic localization of ANP receptors in rat kidney and adrenal gland. Am J Physiol 250:F753–780
Chester AH, Dashwood MR, Clarke JG, Larkin SW, Davies GJ, Tadjkarimi S, Maseri A, Yacoub MH (1989) Influence of endothelin on human coronary arteries and localization of its binding sites. Am J Cardiol 63:1395–1398
Clarke JF, Benjamin N, Larkin SW, Webb DJ, Davies GJ, Maseri A (1989) Endothelin is a potent long-lasting vasoconstrictor in men. Am J Physiol 257:H2033–H2035
Clozel JP, Clozel M (1989) Effects of endothelin on the coronary vascular bed in open-chest dogs. Circ Res 65:1193–1200
Clozel M (1989) Endothelin sensitivity and receptor binding in the aorta of spontaneously hypertensive rats. J Hypertens 7:913–917
Clozel M, Fischli W (1989) Human cultured endothelial cells do secrete endothelin-1. J Cardiovasc Pharmacol 13 (Suppl 5):S229–S231
Cocks TM, Faulkner NL, Sudhir K, Angus J (1989) Reactivity of endothelin-1 on human and canine large veins compared with large arteries in vitro. Eur J Pharmacol 171:17–24
Cohen AJ, Fray JCS (1982) Calçium ion dependence on myogenic renal plasma flow autoregulation: evidence from the isolated perfused rat kidney. J Physiol (Lond) 330:449–460
Criscione L, Thomann H, Rodrigues C, Egleme C, Chiesi M, (1989) Blockade of endothelin-induced contraction by dichlorobenzamil – mode of action. Biochem Biophys Res Commun 163:247–254
Davenport AP, Kaumann AJ, Hall JA, Nunez DJ, Brown MJ (1989a) ^{125}[I] endothelin-binding in mammalian tissue:relation to human atrial inotropic effects and coronary contraction. Brit J Pharmacol 96:135P
Davenport AP, Nunez DJ, Hall JA, Kaumann AJ, Brown MJ (1989b) Autoradiographical localization of binding sites for porcine [^{125}I]endothelin-1 in humans, pigs and rats: functional relevance in humans. J Cardiovasc Pharmacol 13 (Suppl 5):S166–S170
de Bold AJ, Borenstein HB, Veress AT, Sonnenberg H (1981) A rapid and potent natriuretic response to intravenous injection of atrial myocardial extracts in rats. Life Sci 28:89–94
de Carvalho, Nigro D, Scivoletto R, Barbeiro HV, de Oliveira MA, de Nucci G, Fortes ZB (1990) Comparison of the effect of endothelin on microvessels and macrovessels in goldblatt II and deoxycortocosterone acetate-salt hypertensive rats. Hypertension 15 (Suppl I):I68–I71
de Nucci G, Thomas R, D'Orleans-Juste P, Antunes E, Walder C, Warner TD, Vane JR (1988) Pressor effects of circulating endothelin are limited by its removal in the pulmonary circulation and by the release of prostacyclin and endothelium-derived relaxing factor. Proc Natl Acad Sci USA 85:9797–9800
D'Orleans-Juste P, de Nucci G, Vane JR (1989) Endothelin-1 contracts isolated vessels independently of dihydropyridine-sensitive Ca^{2+} channel activation. Eur J Pharmacol 165:289–295
Edwards BS, Zimmerman RS, Burnett JC Jr (1987) Atrial natriuretic factor: physiological actions and implications in congestive heart failure. Cardiovasc Drugs Ther 1:89–100
Eglen RM, Michel AD, Sharif NA, Swank SR, Whiting RL (1989) The pharmacological properties of the peptide, endothelin. Brit J Pharmacol 97:1297–1307
Emori T, Hirata Y, Ohta K, Shichiri M, Marumo F (1989a) Secretory mechanism of immunoreactive endothelin in cultured bovine endothelial cells. Biochem Biophys Res Commun 160:93–100

Emori T, Hirata Y, Ohta K, Shichiri M, Shimokado, K Marumo F (1989b) Concomitant secretion of big endothelin and its C terminal fragment from human and bovine endothelial cells. Biochem Biophys Res Commun 162:217–223

Ezra D, Goldstein RE, Czaja JF, Feuerstein GZ (1989) Lethal ischemia due to intracoronary endothelin in pigs. Am J Physiol 257:H339–H343

Fenwick EM, Marty A, Neher E (1982) Sodium and calcium channels in bovine chromaffin cells. J Physiol 331:599–635

Feraci FM (1989) Effects of endothelin and vasopressin on cerebral blood flow. Am J Physiol 257:H799–H803

Firth JD, Raine AEG, Ratcliffe JGG, Ledingham JG (1988) Endothelin: an important factor in acute renal failure. Lancet ii:1179–1181

Fortes ZB, Scivoletto R, Garcia-Leme J (1989) Endothelin-1 induces potent constriction of lymphatic vessels in situ. Eur J Pharmacol 170:69–73

Fu T, Chang W, Ishida N, Saida K, Mitsui Y, Okano Y, Nozawa Y (1989) Effects of vasoactive intestinal contractor (VIC) and endothelin on intracellular calcium level in neuroblastoma NG 108-15 cells. FEBS Lett 257:351–353

Fukuda Y, Hirata Y, Yoshimi H, Kojima T, Kobayashi Y, Yanagisawa M, Masaki T (1988) Endothelin is a potent secretagogue for atrial natriuretic peptide in cultured rat atrial myocytes. Biochem Biophys Res Commun 155:167–172

Fukuda K, Hori S, Kusuhara M, Satoh T, Kyotani S, Handa S, Nakamura Y, Oono H, Yamaguchi K (1989) Effect of endothelin as a coronary vasoconstrictor in the Langendorff-perfused rat heart. Eur J Pharmacol 165:301–304

Furchgott RF, Zawadzki, JV (1980) The obligatory role of endothelial cells in the relaxation of arterial smooth muscle by acetylcholine. Nature (Lond) 288:373–376

Furlong B, Henderson AH, Lewis MJ, Smith JA (1987) Endothelium-derived relaxing factor inhibits in vitro platelet aggregation. Brit J Pharmacol 90:687–692

Fuxe K, Cintra A, Andbjer B, Anggard E, Goldstein M, Agnati LF (1989) Centrally administered endothelin-1 produces lesions in the brain in the male rat. Acta Physiol Scand 137:155–156

Fyhrquist F, Saijonmaa O, Metsarinne K, Tikkanen I (1989) Presence of endothelin (EN) in human blood and its release following cold provocation. Fourth European Meeting on Hypertension: Abstract No. 275

Galron R, Kloog Y, Bdolah A, Sokolovsky M (1989) Functional endothelin/sarafotoxin receptors in rat heart myocytes: structure – activity relationships and receptor subtypes. Biochem Biophys Res Commun 163:936–943

Gardiner SM, Compton AM, Bennett T, (1989) Regional hemodynamic effects of endothelin-1 in conscious, unrestrained Wistar rats. J Cardiovasc Pharmacol 13 (Suppl 5):S202–S204

Gardner JP, Maher E, Aviv A (1989) Calcium mobilization and Na^+/H^+ antiport activation by endothelin in human skin fibroblasts. FEBS Lett 256:38–42

Geppetti P, Patacchini R, Meini S, Manzini S (1989) Contractile effect of endothelin on isolated iris sphincter muscle of the pig. Eur J Pharmacol 168:117–121

Giaid A, Gibson SJ, Nassif B, Ibrahim N, Legon S, Bloom SR, Yanagisawa M, Masaki T, Varndell JM, Polak JM (1989) Endothelin-1, an endothelium-derived peptide, is expressed in neurons of the human spinal cord and dorsal root ganglia. Proc Natl Acad Sci (USA) 86:7634–7638

Gillespie MN, Owasoyo JO, McMurtry IF, O'Brien RF (1986) Sustained coronary vasoconstriction provoked by a peptidergic substance released from endothelial cells in culture. J Pharmacol Exp Ther 236:339–343

Goetz KL, Wang BC, Madwell JB, Zhu JL, Leadley RJ Jr (1988) Cardiovascular, renal and endocrine responses to intravenous endothelin in conscious dogs. Am J Physiol 255:R1064–1068

Griendling KK, Tsuda T, Alexander RW (1989) Endothelin stimulates diacylglycerol accumulation and activates protein kinase C in cultured vascular smooth muscle cells. J Biol Chem 264:8237–8240

Gu XH, Casley DJ, Nayler WG (1989a) Sarafotoxin S6b displaces specifically bound ^{125}I-endothelin. Eur J Pharmacol 162:509–510

Gu XH, Liu JJ, Dillon JS, Nayler WG (1989b) The failure of endothelin to displace radioactively-labelled calcium antagonists (PN200-110, D888 and diltiazem). Brit J Pharmacol, 96:262–264

Gu XH, Casley DJ, Nayler WG (1989c) Characterization of [^{125}I]endothelin-1 binding sites in rat cardiac membrane fragments. J Cardiovasc Pharmacol 13 (Suppl 5):S171–S173
Gu XH, Casley DJ, Cincotta M, Nayler WG (1990) ^{125}I-endothelin1 binding to brain and cardiac membranes from normotensive and spontaneously hypertensive rats. Eur J Pharmacol 177:205–209
Hamilton MG, Frew R, Lundy PM (1989) Effect of endothelin on Ca^{2+} influx, intracellular free Ca^{2+} levels and ligand binding to N and L type Ca^{2+} channels in rat brain. Biochem Biophys Res Commun 162:1332–1338
Hamilton SL, Yatani A, Hawkes MJ, Redding K, Brown AM (1985) Atrotoxin: a specific agonist for calcium currents in heart. Science 229:182–184
Han SP, Trapani AJ, Fok KF, Westfall TC, Knuepfer MM (1989) Effects of endothelin on regional hemodynamics in conscious rats. Eur J Pharmacol 159:303–305
Hartter E, Woloszczuk W (1989) Radioimmunoassay of endothelin. Lancet i:909
Hartter E, Petzl DH, Schuller W, Woloszcuk W (1989) Low amounts of endothelin in the circulation of healthy persons at rest and at physical stress. Am J Hypertens 2:83A
Hay DWP (1989) Guinea pig tracheal epithelium and endothelin. Eur J Pharmacol 171:241–245
Henry PD (1985) Atherosclerosis, calcium and calcium antagonists. Circulation 72:456–459
Hickey KA, Rubanyi G, Paul RJ, Highsmith RF (1985) Characterization of a coronary vasoconstrictor produced by cultured endothelial cells. Am J Physiol 248:C550–C556
Hiley CR, Pelton JT, Miller RC (1989) Effect of endothelin on field stimulated rat vas deferens and guinea pig ileum. Brit J Pharmacol 96:104P
Hirata Y, Yoshimi H, Takaichi S, Yanagisawa M, Masaki T (1988) Binding and receptor down-regulation of a novel vasoconstrictor endothelin in cultured rat vascular smooth muscle cells. FEBS Lett 239:13–17
Hirata Y (1989) Endothelin-1 receptors in cultured vascular smooth muscle cells and cardiocytes rats. J Cardiovasc Pharmacol 13 (Suppl 5):S157–S158
Hirata Y, Fukuda Y, Yoshima H, Emori T, Shichira M, Marumo F (1989a) Specific receptors for endothelin in cultured rat cardiocytes. Biochem Biophys Res Commun 160:1438–1444
Hirata Y, Itoh KI, Ando K, Endo M, Marumo F, (1989b) Plasma endothelin levels during surgery. N Eng J Med 321:1686
Hirata Y, Matsuoka H, Kimura K, Fukui K, Hayakawa H, Suzuki E, Sugimoto T, Sugimoto T, Yanagisawa M, Masaki T (1989c) Renal vasoconstriction by the endothelial cell-derived peptide endothelin in spontaneously hypertensive rats. Circ Res 65: 1370–1379
Hirata Y, Takagi Y, Fukuda Y, Marumo F (1989d) Endothelin is a potent mitogen for rat vascular smooth muscle cells. Atherosclerosis 78:225–228
Hirata Y, Yoshimi H, Emori T, Shichiri M, Marumo F, Watanabe TX, Kumugaye S, Nakajima K, Kimura T, Sakakibara S, (1989e) Receptor binding activity and cytosolic free calcium response by synthetic endothelin analogues in cultured rat vascular smooth muscle cells. Biochem Biophys Res Commun 160:228–234
Hirata Y, Kanno K, Watanabe TX, Kumagaye S, Nakajima K, Kimura T, Sakakibara S, Marumo F (1990) Receptor binding and vasoconstrictor activity of big endothelin. Eur J Pharmacol 176:225–228
Hoffman A, Keiser HR, Grossman E, Goldstein DS, Gold PW, Kling M (1989) Endothelin concentrations in cerebrospinal fluid in depressive patients. Lancet ii:1519
Hoyer D, Waeber C, Palacois JM (1989) [^{125}I]endothelin-1 binding sites: autoradiographic studies in the brain and periphery of various species, including humans. J Cardiovasc Pharmacol. 13 (Suppl 5):S162–S165
Hu JR, Von Harsdorf R, Lang RE (1988a) Endothelin has potent inotropic effects in rat atria. Eur J Pharmacol 158:275–278
Hu JR, Berninger UC, Lang RE (1988b) Endothelin stimulates atrial natriuretic peptide (ANP) release from rat atria. Eur J Pharmacol 164:177–178
Inoue A, Yanagisawa M, Kimura S, Kasuya Y, Miyauchi T, Goto K, Masaki T (1989a) The human endothelin family: three structurally and pharmacologically distinct isopeptides predicted by three separate genes. Proc Natl Acad Sci (USA) 86:2863–2867
Inoue A, Yanagisawa M, Takuwa Y, Mitsui Y, Kobayashi M, Masaki T (1989b) The human preproendothelin-1 gene. J Biol Chem 264:14954–14959

Ishikawa T, Yanagisawa M, Kimura S, Goto K, Masaki T (1988a) Positive inotropic action of novel vasoconstrictor peptide endothelin on guinea pig atria. Am J Physiol 255:H970–H973

Ishikawa T, Yanagisawa M, Kimura S, Goto K, Masaki T (1988b) Positive chronotropic effects of endothelin, a novel endothelium-derived vasoconstrictor peptide. Pflügers Arch 413:108–110

Itoh Y, Yanagisawa M, Ohkubo S, Kimura C, Kosaka T, Inoue A, Ishida N, Mitsui Y, Onda H, Fujino M, Masaki T (1988) Cloning and sequence analysis of cDNA encoding the precursor of a human endothelium-derived vasoconstrictor peptide, endothelin: identity of human and porcine endothelin. FEBS Lett 231:440–444

Itoh Y, Kimura C, Onda H, Fujino M (1989) Canine endothelin-2 cDNA for mature polypeptide. Nucleic Acids Research 17:5389

Jones CR, Pelton JT, Miller RC, Hiley CR (1989) Binding of [^{125}I]endothelin to rat brain and its inhibition by peptide analogues. Brit J Pharmacol 96:Abstract 346P

Jones CR, Hiley CR, Pelton, JT Mohr M (1989a) Autoradiographic visualization of the binding site for ^{125}I-endothelin in rat and human brain. Neurosciences Letters 97:276–279

Jones CR, Hiley CR, Pelton, JT, Miller, RC (1989b) Autoradiographic localization of endothelin binding sites in kidney. Eur J Pharmacol 163:379–382

Kai H, Kanaide H, Nakamura M (1989) Endothelin-sensitive intracellular Ca^{2+} store overlaps with caffeine-sensitive one in rat aortic smooth muscle cells in primary culture. Biochem Biophys Res Commun 158:235–243

Kanse SM, Ghatei MA, Bloom SR (1989) Endothelin binding sites in porcine aortic and rat lung membranes. Eur J Biochem 182:175–179

Karwatowska-Prokopczuk E, Wennmalm A (1990) Effects of endothelin on coronary flow, mechanical performance, oxygen uptake, and formation of purines and on outflow of prostacyclin in the isolated rabbit heart. Circ Res 66:46–54

Kasuya Y, Ishikawa T, Yanagisawa M, Kimura S, Goto K, Masaki T (1989a) Mechanism of contraction to endothelin in isolated porcine coronary artery. Am J Physiol 257:H1828–H1835

Kasuya Y, Takuwa Y, Yanagisawa M, Kimura S, Goto K, Masaki T (1989b) Endothelin-1 induces vasoconstriction through two functionally distinct pathways in porcine coronary artery: contribution of the phosphoinositide turnover. Biochem Biophys Res Commun 161:1049–1055

Kauser K, Rubanyi GM, Harder DR (1990) Endothelium-dependent modulation of endothelial-induced vasoconstriction and membrane depolarization in cat cerebral arteries. J Pharmacol Exp Ther 252:93–97

Kawaguchi H, Sawa H, Yasuda H (1990) Endothelin stimulates angiotensin I to angiotensin II conversion in cultured pulmonary artery endothelial cells. J Mol Cell Cardiol (in press)

Kimura S, Kasuya Y, Sawamura T, Shinmi O, Sugita Y, Yanagisawa M, Goto K, Masaki T (1989) Conversion of big endothelin-1 to 21 residue endothelin-1 is essential for expression of full vasoconstrictor activity: structure activity relationships of big endothelin-1. J Cardiovasc Pharmacol 13 (Suppl 5):S5–S7

King AJ, Brenner BM, Anderson S (1989) Endothelin: a potent renal and systemic vasoconstrictor peptide. Am J Physiol 256:F1051–F1058

Kitamura K, Tanaka T, Kato J, Eto T, Tanaka K (1989a) Regional distribution of immunoreactive endothelin in porcine tissue: abundance in inner medulla of kidney. Biochem Biophys Res Commun 161:348–352

Kitamura K, Tanaka T, Kato J, Ogawa T, Eto T, Tanaka K, (1989b) Immunoreactive endothelin in rat kidney inner medulla: marked decrease in spontaneously hypertensive rats. Biochem Biophys Res Commun 162:38–44

Kloog Y, Ambar I, Sokolovsky M, Kochva E, Wollberg Z, Bdolah A (1988) Sarafotoxin, a novel vasoconstrictor peptide: phosphoinositide hydrolysis in rat heart and brain. Science 242:268–270

Kochva E, Viljoen CC, Botes DP (1982) A new type of toxin in the venom of snakes of the genus *Atractaspis* (Atractaspidinae). Toxicon 20:581–592

Kohzuki M, Johnston CI, Chai SY, Casley DJ, Mendelsohn FAO (1989) Localization of endothelin-receptors in rat kidney. Eur J Pharmacol 160:193–194

Komuro I, Kurihara H, Sugiyama T, Takuku F, Yazaki Y (1988a) Endothelin stimulates *c-fos* and *c-myc* expression and proliferation of vascular smooth muscle cells. FEBS Lett 238:249–252

Komuro I, Kurihara H, Sugiyama T, Yoshizumi M, Yazaki Y (1988b) Endothelin, endothelial cell-derived potent vasoconstrictor peptide, stimulates c-fos and c-myc expression and proliferation of vascular smooth muscle cells. Circulation 78 (Suppl 2):S394 (abstract) 1572

Korbut R, Lidbury P, Thomas GR, Vane JR (1989) Fibrinolytic activity of endothelin-1. Thrombosis Research 55:797–799

Kosaka T, Suzuki N, Matsumoto H, Itoh Y, Yasuhara T, Onda H, Fujino M (1989) Synthesis of the vasoconstrictor peptide endothelin in kidney cells. FEBS Lett 249:42–46

Koseki C, Imai M, Hirata Y, Yanagisawa M, Masaki T (1989a) Autoradiographic distribution in rat tissues of binding sites for endothelin: a neuropeptide. Am J Physiol 256:R858–R866

Koseki C, Imai M, Hirata Y, Yanagisawa M, Masaki T, (1989b) Binding sites for endothelin-1 in rat tissues: an autoradiographic study. J Cardiovasc Pharmacol 13 (suppl 5): S153–S155

Koyama H, Tabata T, Nishzawa Y, Inoue T, Mori H, Yamaji T (1989) Plasma endothelin levels in patients with uraemia. Lancet i:991

Kozuka M, Ito T, Hirose S, Takahashi K, Hagiwara H (1989) Endothelin – induces two types of contractions of rat uterus: phasic contractions by way of voltage-dependent calcium channels and developing contractions through a second type of calcium channels. Biochem Biophys Res Commun 159:317–323

Kurihara H, Yamaoki K, Nagai R, Yoshizumi M, Takaku F, Satoh H, Inui J (1989) Endothelin: a potent vasoconstrictor associated with coronary vasospasm. Life Sci 44:1937–1943

Kurihara H, Yamaoki K, Nagai R, Yoshizumi M, Takaku F, Satoh H, Inui J, Yasaki Y (1989a) Coronary vasoconstriction produced by endothelin, a novel endothelium-derived vasoconstrictor peptide. Circulation 78 (Suppl II):171

Kurihara H, Yoshizumi M, Sugiyama T, Yamaoki K, Nagai R, Takaku F, Satoh H, Inui J, Yanagisawa M, Masaki T, Yazaki Y (1989b) The possible role of endothelin-1 in the pathogenesis of coronary vasospasm. J Cardiovasc Pharmacol 13 (Suppl 5):S132–S137

Kusuhara M, Yamaguchi K, Ohnishi A, Abe K, Kimura S, Oono H, Hori S, Nakamura Y (1989) Endothelin potentiates growth-factor-stimulated DNA synthesis in Swiss 3T3 cells. Jap J of Cancer Res 80:302–305

Lagente V, Chabrier PE, Mencia-Huerta JM, Braquet P (1989) Pharmacological modulation of the bronchopulmonary action of the vasoactive peptide, endothelin, administered by aerosol in the guinea pig. Biochem Biophys Res Commun 158:625–632

Lam HC, Takahashi K, Ghatei MA, Bloom SR (1990) Presence of immunoreactive endothelin in human milk. FEBS Lett 261:184–186

Lazdunski M (1983) Apamin, a neurotoxin specific for one class of Ca^{2+}-dependent K^+ channels. Cell Calcium 4:421–428

Lee CY, Chiappinelli VA (1988) Similarity of endothelin to snake venom toxin. Nature (Lond) 335:303

Lee KH, Leikauf GD, Sperelakis N (1990) Electromechanical effects of endothelin on ferret bronchial and tracheal smooth muscle. Am J Physiol 68:417–420

Lee TS, Chao T, Hu QK, King GL (1989) Endothelin stimulates a sustained 1,2-diacylglycerol increase and protein kinase C activation in bovine aortic smooth muscle cells. Biochem Biophys Res Commun 162:381–386

Lichtlen PR, Hugenholtz PG, Rafflenbeul W, Hecker H, Jost S, Deckers JW (1990) Retardation of the angiographic progression of coronary artery disease in many by the calcium channel blocker nifedipine – results of the International Nifedipine Trial on Antiatherosclerotic Therapy (Intact). Lancet 335:109–113

Lidbury PS, Thiermermann C, Thomas GR, Vane JR (1989) Endothelin-3: selectivity as an anti-aggregatory peptide in vivo. Eur J Pharmacol 166:335–338

Linz W, Albus U, Schilkens (1989) Effects of endothelin on isolated ischaemic rat hearts during ramiprilat, bradykinin and indomethacin perfusion. J Hypertens 7 (Suppl 6):S310S311

Lippton H, Goff J, Hymann A (1988) Effects of endothelin in the systemic and renal vascular beds in vivo. Eur J Pharmacol 155:197–199

Liu JJ, Casley DJ, Nayler WG (1989) Ischaemia causes externalization of endothelin-1 binding sites in rat cardiac membranes. Biochem Biophys Res Commun 164:1220–1225

Liu JJ, Chen DJ, Casley DJ, Nayler WG (1990a) Effect of ischaemia and reperfusion on ^{125}I endothelin binding in rat cardiac membranes. Am J Physiol. 258:H829–H835

Liu JJ, Kelly J, Casley DJ, Johnston CI, Nayler WG (1990b) Solubilization and characterization of endothelin-1 receptors in rat cardiac tissue. Biochem Biophys Res Commun 166:299–307

Lopez-Farré A, Montañés I, Millas I, Lopez-Novoa JM (1989) Effect of endothelin on renal function in rats. Eur J Pharmacol 163:187–189

Loutzenhiser R, Epstein M, Hayashi K, Horton C (1990) Direct visualization of effects of endothelin on the renal vasculature. Am J Physiol 258:F61–F68

Lowenstein J, Beranbaum ER, Chasis H, Baldwin DS (1970) Intrarenal pressure and exaggerated natriuresis in essential hypertension. Clin Sci 38:359–374

MacCumber MW, Ross CA, Glaser BM, Snyder SH (1989) Endothelin: Visualization of mRNAs by *in situ* hybridization provides evidence for local action. Proc Natl Acad Sci (USA) 86:7285–7289

McMahon EG, Fox KF, Moore WM, Smith CE, Siegel NR, Trapani AJ (1989) In vitro and in vivo activity of chymotrypsin-activated big endothelin (porcine 1–40) Biochem Biophys Res Commun 161:406–413

Madeddu P, Glorioso N (1989) Effects of endothelin on blood pressure and renal function in awake, normotensive rats. Am J Hypertens 2:Abstract 1473

Maggi CA, Patacchini R, Giuliani S, Meli A (1989a) Potent contractile effect of endothelin in isolated guinea-pig airways. Eur J Pharmacol 160:179–182

Maggi CA, Giuliani S, Patacchini R, Santicioli P, Rovero P, Giachetti A, Meli A (1989b) The C-terminal hexapeptide, endothelin-(16–21) discriminates between different endothelin receptors. Eur J Pharmacol 166:121–122

Maggi CA, Giuliani S, Patacchini R, Rovero P, Giachetti A, Meli A (1989c) The activity of peptides of the endothelin family in various mammalian smooth muscle preparations. Eur J Pharmacol 174:23–31

Maggi CA, Giuliani S, Patacchini R, Santicioli P, Giachetti A, Meli A (1990) Further studies on the response of the guinea pig isolated bronchus to endothelins and sarafotoxin S6b. Eur J Pharmacol 176:1–9

Marche P (1989) Membrane phosphoinositide metabolism in hypertension. News in Physiol Sci 4:230–233

Marsden PA, Danthuluri NR, Brenner BM, Ballermann BJ, Brock TA (1989a) Endothelin action on vascular smooth muscle involves inositol triphosphate and calcium mobilization. Biochem Biophys Res Commun 158:86–92

Marsden PA, Martin ER, Dorfman D, Brock TA, Brenner BM, Collins T, Ballerman BJ (1989b) Endothelin: Gene expression, release and action in cultured cells of the renal glomerulus. Am J Hypertens 2:49A

Masaoka H, Suzuki R, Hirata Y, Emori T, Marumo F, Hirakawa K (1989) Raised plasma endothelin in aneurysmal subarachnoid haemorrhage. Lancet ii:1402

Masuda Y, Miyazaki H, Kondoh M, Watanabe H, Yanagisawa M, Masaki T, Murakami K (1989) Two different forms of endothelin receptors in rat lung. FEBS Lett 257:208–210

Matsumoto T, Suzuki N, Onda H, Fujino M (1989) Abundance of endothelin-3 in rat intestine, pituitary gland and brain. Biochem Biophys Res Commun 164:74–80

Matsumura Y, Nakase K, Ikegawa R, Hayashi K, Ohyama T, Morimoto S (1989) The endothelium-derived inhibits renin release in vitro. Life Sci 44:149–158

Mendelsohn FAO, Allen AM, Chai SY, Sexton PM, Figdor R (1987) Overlapping distributions of receptors for atrial natriuretic peptide and angiotensin 11 visualized by in vitro autoradiography: morphological basis of physiological antagonism. Can J Physiol Pharmacol 65:1517–1521

Mené P, Simonson MS, Dunn MJ (1989) Physiology of the mesangium. Physiol Rev 69:1347–1424

Meulemans AL, Brutsaert DL (1988) Does endocardial endothelium mediate positive inotropic response to angiotensin I and II? Eur Heart J 9 (Suppl 1):3–7

Meyer-Lehnert H, Wanning C, Backer A, Stelkens H, Kramer HJ (1989a) Transient and sustained increase by endothelin of intracellular calcium $[Ca^{2+}]_i$ in vascular smooth muscle cells (VSMC): relation to inhibition of Na-K-ATPase? Am J Hypertens 2:Abstract 1049

Meyer-Lehnert H, Wanning C, Predel HG, Backer A, Stelkens H, Kramer HJ (1989b) Effects of endothelin on sodium transport mechanisms; potential role of cellular Ca^{2+} mobilization. Biochem Biophys Res Commun 163:458–465

Miller RJ (1987) Multiple calcium channels and neuronal function. Science 235, 46–52

Miller VM, Komori K, VanHoutte PM (1989a) Response of canine veins to endothelin. J Vasc Med Biol 1(2): Abstract 104

Miller WL, Redfield MM, Burnet C (1988) Integrated cardiac, renal and endocrine actions of endothelin. J Clin Invest 83: 317–320

Miller WL, Cavero PG, Aarhus LL, Heublein DM, Burnett JC Jr (1989) Endothelin-induced vasoconstriction and catecholamine release are attenuated by nitroglycerine. Am Heart J 2:37A

Minkes RK, Coy DH, Murphy WA, McNamara DB, Kadowitz PJ (1989) Effects of porcine and rat endothelin and an analog on blood pressure in the anesthetized cat. Eur J Pharmacol 164:571–575

Mir AK, Berthold H, Schoitysik GE, Fozard JR (1989) Cardiovascular effects of endothelin-1 in pithed spontaneously hypertensive rats: evaluation of its mechanisms of action. Naunyn-Schmiedeberg's Arch Pharmacol 340:424–430

Mitsuhashi T, Morris RC, Ives HE (1989) Endothelin-induced increases in vascular smooth muscle Ca^{2+} do not depend on dihydropyridine-sensitive Ca^{2+} channels. J Clin Invest 84:635–639

Miura K, Yukimura T, Yamashita Y, Shimmen T, Okumura M, Imanishi M, Yamamoto K (1989) Endothelin stimulates the renal production of prostaglandin E_2 and I_2 in anaesthetized dogs. Eur J Pharmacol 170:91–93

Miyamori I, Matsubara T, Ito Y, Koshida H, Takeda R (1989) Cardiovascular, renal and regional effects of endothelin (ET) in rabbits: effect of an ET antibody. Am J Hypertens 2:87A (Abstract 1347)

Miyauchi T, Yanagisawa M, Tomizawa T, Sugishita Y, Suzuki N, Fujino M, Ajisaka R, Goto K, Masaki T (1989a) Increased plasma concentrations of endothelin-1 and big endothelin-1 in acute myocardial infarction. Lancet i:53

Miyauchi T, Ishikawa T, Tomobe Y, Yanagisawa M, Kimura S, Sugishita Y, Ito I, Goto K, Masaki T (1989b) Characteristics of pressor response to endothelin in spontaneously hypertensive and Wistar-Kyoto rats. Hypertension 14:427–434

Miyazaki H, Kondoh M, Watanabe H, Masuda Y, Murakami K, Takahashi M, Yanagisawa M, Kimura S, Goto K, Masaki T (1990) Affinity labelling of endothelin receptor and characterization of solubilized endothelin - endothelin-receptor complex. Eur J Biochem 187:125–129

Moncada S, Gryglewski R, Bunting S, Vane JR (1976) An enzyme isolated from arteries transforms prostaglandin endoperoxidase to an unstable substance that inhibits platelet aggregation. Nature (Lond) 263:665–666

Moravec CS, Reynolds EE, Stewart RW, Bond M (1989) Endothelin is a positive inotropic agent in human and rat heart in vitro. Biochem Biophys Res Commun 159:14–18

Morel DR, Lacroix JS, Hemsen A, Steinig DA, Pittet JF, Lundber JM (1989) Increased plasma and pulmonary lymph levels of endothelin during endotoxin shock. Eur J Pharmacol 167:427–428

Moser PC, Pelton JD (1989) Behaviourial effects of centrally administered endothelin in the rat. Brit J Pharmacol 96: Abstract 347P

Muldoon LL, Rodland KD, Forsythe ML, Magun BE (1989) Stimulation of phosphatidylinositol hydrolysis, diacylglycerol release and gene expression in response to endothelin, a potent new agonist for fibroblasts and smooth muscle cells. J Biol Chem 264:8529–8536

Munro SL, Craik DJ, McConville D, Hall JG, Searle M, Bicknell W, Sclanlon D, Chandler CJ (1990) Solution conformation of endothelin by ^1H NMR spectroscopy and distance geometry. FEBS Lett (submitted)

Nakaki T, Nakayama M, Yamomoto S, Kato R (1989) Endothelin-mediated stimulation of DNA synthesis in vascular smooth muscle cells. Biochem Biophys Res Commun 158:880–883

Nakajima K, Kumagaye S, Nishio H, Kuroda H, Watanabe T, Kobayashi Y, Tamaoki H, Kimura T, Sakakibara S (1989a) Synthesis of endothelin-1 analogues, endothelin-3, and sarafotoxin S6b: Structure activity relationships. J Cardiovasc Pharmacol 13 (Suppl 5):S8–S12

Nakajima K, Kubo S, Kumagaye S, Nishio H, Tsunemi M, Inui T, Kuroda H, Chino N, Watanabe TX, Kimura T, Sakakibara S (1989b) Structure-activity relationship of endothelin: importance of charged groups. Biochem Biophys Res Commun 163:424–429

Nakajo S, Sugiura M, Snajdar RM, Boehm FH, Inagami T (1989) Solubilization and identification of human placental endothelin receptor. Biochem Biophys Res Commun 164:205–211

Nakamoto H, Suzuki H, Murakami M, Kageyama Y, Ohishi A, Fukuda K, Hori S, Saruta T (1989) Effects of endothelin on systemic and renal haemodynamics and neuroendocrine hormones in conscious dogs. Clin Sci 77:567–572

Naruse M, Tsuchiya K, Sanaka T, Naruse K, Zeng ZP, Nitta K, Sugino N, Shizume K (1989) Effects of endothelin on systemic haemodynamics and renal functions in anaesthetized dogs. Am J Hypertens 2:58A (Abstract 1230)

Nayler WG (1983) Calcium and Cell Death. Eur Heart J 4:33–41

Nayler WG (1988) Ion-conducting calcium channels. In: Nayler WG (ed) The calcium antagonists. Academic Press, London, pp23–44

Nayler WG (1989) Endothelin-1 and the cardiovascular system: a mini-review. J Appl Cardiol 4:495–503

Nayler WG, Gu XH, Casley DJ, (1989a) Sarafotoxin S6c is a relatively weak displacer of specifically bound ^{125}I endothelin. Biochem Biophys Res Commun 161:89–94

Nayler WG, Gu XH, Casley DJ, Panagiotopoulos S, Liu J, Mottram PL (1989b) Cyclosporine increases endothelin-1 binding site density in cardiac cell membranes. Biochem Biophys Res Commun 163:1270–1294

Nayler WG, Liu J, Panagiotopoulos S, Casley DJ (1989c) Streptozotocin-induced diabetes reduces the density of endothelin-1 binding sites in rat cardiac membranes. Brit J Pharmacol 97:993–995

Nayler WG (1990) Endothelin:isoforms, binding sites and possible implications in pathology. Trends Pharm Sci 11:96–99

Neuser D, Zaiss S, Stasch JP (1990) Endothelin receptors in cultured renal epithelial cells. Eur J Pharmacol 176:241–243

Nomura A, Uchida Y, Kameyama M, Saotome M, Oki K, Hasegawa S (1989) Endothelin in bronchial asthma. Lancet ii:747

O'Brien RF, McMurtry IF (1984) Endothelial cell supernatants contract bovine pulmonary artery rings. Am Rev of Respir Dis 129:A337

O'Brien RF, Robbins RJ, McMurtry IF (1987) Endothelial cells in culture produce a vasoconstrictor substance. J Cell Physiol 132:263–270

Ohkubo S, Itoh Y, Kimura C, Onda H, Fujino M (1990) Nucleoside sequence of a rabbit genomic DNA encoding mature endothelin-3. Nucleic Acids Research 18:374

Ohlstein EH, Horohonich S, Hay DWP (1989) Cellular mechanisms of endothelin in rabbit aorta. J Phamacol Exp Ther 250:548–555

Ohman KP, Hoffman A, Keiser HR (1989) Endothelin is a potent regulator of arterial pressure with effects on the kidney and on atrial natriuretic peptide levels. Am J Hypertens 2:Abstract 1071

Ohnishi A, Yamaguchi K, Kusuhara M, Abe K, Kimura S (1989) Mobilization of intracellular calcium by endothelin in Swiss 3T3 cells. Biochem Biophys Res Commun 161:489–495

Orita Y, Fujiwara Y, Ochi S, Takami T, Fukunaga M, Yokoyama K (1989) Endothelin-1 receptors in rat renal glomeruli. J Cardiovasc Pharmacol 13 (Suppl 5):S159–S161

Ouchi Y, Seungbum K, Souza AC, Iijima S, Hattori A, Orima H, Yoshizumi M, Kurihara H, Yazaki Y (1989) Central effect of endothelin on blood pressure of conscious rats. Am J Physiol 256:H1747–1751

Pang DC, Johns A, Patterson K, Parker-Bothelho LH, Rubanyi GM (1989) Endothelin-1 stimulates phosphatidylinositol hydrolysis and calcium uptake in isolated canine coronary arteries. J Cardiovasc Pharmacol 13 (Suppl 5):S75–S79

Pernow J, Hemsen A, Lundberg JM (1989a) Increased plasma levels of endothelin-like immunoreactivity during endotoxin administration in the pig. Acta Physiol Scand 137:317–318

Pernow J, Hemsen A, Lundberg JM (1989b) Tissue specific distribution, clearance and vascular effects of endothelin in the pig. Biochem Biophys Res Commun 161:647–653

Pernow J, Matran R, Lundberg JM (1989c) Effect of endothelin-1 on regional vascular resistance in pigs. J Cardiovasc Pharmacol 13 (Suppl 5):S205–S206)

Power RF, Wharton J, Zhao Y, Bloom SR, Polak JM (1989a) Autoradiographic localization of endothelin-1 binding sites in the cardiovascular and respiratory systems. J Cardiovasc Pharmacol 13 (Suppl 5):S50–S56

Power RF, Wharton J, Salas SP, Kanse S, Ghatei M, Bloom SR, Polak JM (1989b) Autoradiographic localisation of endothelin binding sites in human and porcine coronary arteries. Eur J Pharmacol 160:199–200

Radomski MW, Palmer RMJ, Moncada S (1987) Endogenous nitric oxide inhibits human platelet adhesion to vascular endothelium. Lancet ii:1057–1058

Rae GA, Trybulec M, de Nucci G, Vane JR (1989) Endothelin-1 releases eicosanoids from rabbit isolated perfused kidney and spleen. J Cardiovasc Pharmacol 13 (Suppl 5):S89–S92

Randall MD, Douglas SA, Hiley CR (1989) Vascular activities of endothelin-1 and some alanyl substituted analogues in resistance beds of the rat. Br J Pharmacol 98:685–699

Reid JJ, Wong-Dusting HK, Rand MJ (1989) The effect of endothelin on noradrenergic transmission in rat and guinea pig atria. Eur J Pharmacol 168:93–96

Resnik TJ, Scott-Burden T, Buhler FR (1989) Endothelin stimulates phospholipase C in cultured vascular smooth muscle cells. Biochem Biophys Res Commun 157:1360–1368

Richards NT, Poston L, Goldsmith DJA, Cragoe EJ, Hilton PJ (1989) Endothelin-induced contraction of human peripheral resistance vessels is partly dependent on stimulation of sodium-hydrogen exchange. J Hypertens 7:770–780

Ross RN (1986) The pathogenesis of atherosclerosis. N Eng J Med 314:488–500

Roubert P, Gillard V, Plas P, Guillon JM, Chabrier PE, Braquet P (1989) Angiotensin 11 and phorbol-esters potently down-regulate endothelin (ET-1) binding sites in vascular smooth muscle cells. Biochem Biophys Res Commun 164:809–815

Roubert P, Gillard V, Plas P, Chabrier PE, Braquet P (1990) Up-regulation of endothelin receptors by cholera toxin in cultured vascular smooth muscle cells. Fifth Scientific Meeting of the American Society of Hypertension: Abstract 1255

Rozengurt E (1986) Early signals in the mitogenic response. Science 234:161–166

Rozengurt N, Springall DR, Polak JM (1990) Localization of endothelin-like immunoreactivity in airway epithelium of rats and mice. J Pathol 160:5–8

Rubanyi GM (1988) Endothelium-dependent pressure-induced contraction of isolated canine carotid arteries. Am J Physiol 255:H783–H788

Ryan US, Glassberg MK, Nolop KB (1989) Endothelin-1 from pulmonary artery and microvessels acts on vascular and airways smooth muscle. J Cardiovasc Pharmacol 13 (Suppl 5):S50–S56

Saida K, Mitsui Y, Ishida N (1989) A novel peptide, vasoactive intestinal contractor, of a new (endothelin) peptide family. J Biol Chem 264:14613–14616

Saito Y, Nakao K, Itoh H, Yamada T, Mukoyama M, Ari H, Hosoda K, Shirakami G, Suga SI, Jougasaki M, Morichika S, Imura H (1989a) Endothelin in human plasma and culture medium of aortic endothelial cells-detection and characterization with radioimmunoassay using monoclonal antibody. Biochem Biophys Res Commun 161:320–326

Saito T, Yanagisawa M, Miyauchi T, Suzuki N, Matsumoto H, Fujino M, Masaki T (1989b) Endothelin in human circulating blood: effects of major surgical stress. Jap J Pharmacol 49:Abstract 215P

Saito Y, Nakao K, Mukoyama M, Imura H (1990) Increased plasma endothelin level in patients with essential hypertension. N Eng J Med 322:205

Sakata K, Ozaki H, Kwon SC, Haraki H (1989) Effects of endothelin on the mechanical activity and cytosolic calcium levels of various types of smooth muscle. Brit J Pharmacol 98:483–492

Saliminen K, Tikkanen I, Saijonmaa O, Nieminen M, Fyhrquist F, Frick MH (1989) Modulation of coronary tone in acute myocardial infarction by endothelin. Lancet ii:747

Sawamura T, Kimura S, Shinmi O, Sugita Y, Yanagisawa M, Masaki T (1989) Analysis of endothelin-related peptides in culture supernatant of porcine aortic endothelial cells: evidence for biosynthetic pathway of endothelin-1. Biochem Biophys Res Commun 162:1287–1294

Schini VB, Hendrickson H, Heublein DM, Burnett JC Jr, Vanhoutte PM (1989) Thrombin enhances the release of endothelin from cultured porcine aortic endothelial cells. Eur J Pharmacol 165:333–334

Schramm M, Thomas G, Towart M, Franckowiak G (1983) Novel dihydropyridines with positive inotropic action through activation of Ca^{2+} channels. Nature (Lond) 303:535–537

Schwartz SM, Campbell GR, Campbell JH (1986) Replication of smooth muscle cells in vascular disease. Circ Res 58: 427–444

Shah AM, Lewis MJ, Henderson AH (1989) Inotropic effects of endothelin in ferret ventricular myocardium. Eur J Pharmacol 163:365–367

Shiba R, Yanagisawa M, Miyauchi T, Ishii Y, Kimura S, Uchiyama Y, Masaki T, Goto K (1989) Elimination of intravenously injected endothelin-1 from the circulation of the rat. J Cardiovasc Pharmacol 13 (Suppl 5):S98–S101

Shichiri M, Hirata Y, Kanno K, Ohta K, Emori T, Marumo F (1989) Effect of endothelin-1 on release of arginine-vasopressin from perfused rat hypothalamus. Biochem Biophys Res Commun 163:1332–1337

Shichiri M, Hirata Y, Ando K, Kanno K, Emori T, Ohta K, Marumo, F (1990) Postural change and volume expansion affect plasma endothelin levels. J Am Med Assoc 263:661

Shigeno T, Mima T, Takakura K, Yanagisawa M, Saito A, Goto K, Masaki T (1989) Endothelin-1 acts on cerebral arteries from the adventitial but not from the luminal side. J Cardiovasc Pharmacol 13 (Suppl 5):S174–S176

Shinmi O, Kimura S, Yoshizawa T, Sawamura T, Uchiyama Y, Sugita Y, Kanazawa I, Yanagisawa M, Goto K, Masaki T (1989a) Presence of endothelin-1 in porcine spinal chord isolation and sequence determination. Biochem Biophys Res Commun 162:340–346

Shinmi O, Kimura S, Sawamura T, Sugita Y, Yoshizawa T, Uchiyama Y, Yanagisawa M, Goto K, Masaki T, Kanazawa I (1989b) Endothelin-3 is a novel neuropeptide: isolation and sequence determination of endothelin-1 and endothelin-3 in porcine brain. Biochem Biophys Res Commun 164:587–593

Silberg SD, Poder TC, Lacerda AC (1989) Endothelin increases single-channel calcium currents in coronary artery smooth muscle cells. FEBS Lett 247:68–72

Simonson MS, Wann S, Mene P, Dubyak GR, Kester M, Nakazato Y, Sedor JR, Dunn MJ (1989) Endothelin stimulates phospholipase C, Na^+/H^+ exchange, C-fos expression, and mitogenesis in rat mesangial cells. J Clin Invest 83:708–712

Simonson MS, Dunn MJ (1990) Endothelin: Pathways in Transmembrane signalling. Hypertension 15 (Suppl 1):1–5

Simpson AWM, Ashley CC (1989) Endothelin-evoked Ca^{2+} transients and oscillations in AIO vascular smooth muscle cells. Biochem Biophys Res Commun 163:1223–1229

Sneddon JM, Vane JR (1988) Endothelium-derived relaxing factor reduces platelet adhesion to bovine endothelial cells. Proc Natl Acad Sci (USA) 85:2800–2804

Stasch JP, Kazda S (1989) Endothelin-1 induced vascular contractions: Interactions with drugs affecting the calcium channel. J Cardiovasc Pharmacol 13 (Suppl 5):S63–S66

Sugiura M, Inagami T, Hare GMT, Johns JA (1989a) Endothelin action: inhibition by a protein kinase C inhibitor and involvement of phosphoinositols. Biochem Biophys Res Commun 159:170–176

Sugiura M, Inagami T, Kon V (1989b) Endotoxin stimulates endothelin-release in vivo and in vitro as determined by radioimmunoassay. Biochem Biophys Res Commun 161:1220–1227

Sugiura M, Snajdav RM, Schwartzberg M, Badr KF, Inagami T (1989c) Identification of two types of specific endothelin receptors in rat mesangial cell. Biochem Biophys Res Commun 162:1396–1401

Sunnergren KP, Word RA, Sambrook JF, MacDonald PC, Casey ML (1990) Expression and regulation of endothelin precursor mRNA in a vascular human amnion. Mol Cell Endocrinol 68:R7–R14

Supattapone S, Simpson AWM, Ashley CC (1989) Free calcium rise and mitogenesis in glial cells caused by endothelin. Biochem Biophys Res Commun 165:1115–1122

Suzuki N, Matsumoto H, Kitada C, Kimura S, Fujino M (1989) Production of endothelin-1 and big endothelin-1 by tumour cells with epithelial-like morphology. J Biochem 106:736–741

Tabuchi Y, Nakamaru M, Ragugi H, Nagano M, Ogihara T (1989) Endothelin enhances adrenergic vasoconstriction in perfused rat mesenteric arteries. Biochem Biophys Res Commun 159:1304–1308

Takahashi K, Brooks RA, Kanse SM, Ghatei MA, Kohner EM, Bloom SR (1989) Production of endothelin-1 by cultured bovine retinal endothelial cells and presence of endothelin receptors on associated pericytes. Diabetes 38:1200–1202

Takasaki C, Tamiya N, Bdolah V, Wollberg Z, Kochva E (1988a) Sarafotoxins S6: several isotoxins from *Atractaspis Engaddensis* (burrowing asp) venom that affect the heart. Toxicon 26:543–548

Takasaki C, Yanagisawa M, Kimura S, Goto K, Masaki T (1988b) Similarity of endothelin to snake venom toxin. Nature (Lond) 335:303

Takaoka M, Takenobu Y, Miyata Y, Ikegawa R, Matsumura Y, Morimoto S (1990) Pepsin, an aspartic protease, converts porcine big endothelin to 21 residue endothelin. Biochem Biophys Res Commun 166:436–442

Takayanagi R, Hashiguchi T, Ohashi M, Nawata H (1990) Regional distribution of endothelin receptor in porcine cardiovascular tissues. Regulatory Peptides 27:247–255

Takuwa N, Takuwa Y, Yanagisawa M, Yamashita K, Masaki T (1988) A novel vasoactive peptide endothelin stimulates mitogenesis through inositol lipid turnover in Swiss 3T3 fibroblasts. J Biol Chem 264:7856–7861

Terashita ZI, Shibouta Y, Imura Y, Iwasaki K, Nishikawa K (1989) Endothelin-induced sudden death and the possible involvement of platelet activating factor (PAF). Life Sci 45:1911–1918

Thiemermann C, Lidbury PS, Thomas GR, Vane JR (1988) Endothelin inhibits ex vivo platelet aggregation in the rabbit. Eur J Pharmacol 158:181–182

Thiermermann C, Lidbury PS, Thomas GR, Vane JR (1989) Endothelin-1 releases prostacyclin and inhibits ex vivo platelet aggregation in the anaesthetized rabbit. J Cardiovasc Pharmacol 13 (Suppl 5):S138–S141

Thiemermann C, May GR, Page CP, Vane JR (1990) Endothelin-1 inhibits platelet aggregation *in vivo:* a study with ^{111}indium-labelled platelets. Brit J Pharmacol 99:303–308

Tippins JR, Antoniw JW, Maseri A (1989) Endothelin-1 is a potent constrictor in conductive and resistive coronary arteries. J Cardiovasc Pharmacol 13 (Suppl 5):S177–S179

Tomita K, Ujiie K, Nakanishi T, Tomura S, Matsuda O, Ando K, Shichiri M, Hirata Y, Marumo F (1989) Plasma endothelin levels in patients with acute renal failure. The N Eng J Med 321:1127

Tomobe Y, Miyauchi T, Saito A, Yanagisawa M, Kimura S, Goto K, Masaki T (1988) Effects of endothelin on the renal artery from spontaneously hypertensive and Wistar Kyoto rats. Eur J Pharmacol 152:373–374

Turner NC, Power RF, Polak JM, Bloom SR, Dollery CT (1988) Contraction of rat tracheal smooth muscle by endothelin. Brit J Pharmacol 96:103P

Turner NC, Power RF, Polak JM, Bloom SR, Dollery CT (1989) Endothelin-induced contractions of tracheal smooth muscle and identification of specific endothelin binding sites in the trachea of the rat. Brit J Pharmacol 98:361–366

Uchida Y, Ninomiya H, Saotome M, Nomura A, Ohtsuka M, Yanagisawa M, Goto K, Masaki T, Hasegawa S (1988) Endothelin, a novel vasoconstrictor peptide as potent bronchoconstrictor. Eur J Pharmacol 154:227–228

Van Renterghem C, Vigne P, Barhanin J, Schmid-Alliana A, Frelin C, Lazdunski M (1988) Molecular mechanisms of action of the vasoconstrictor peptide, endothelin. Biochem Biophys Res Commun 157:977–985

Vigne P, Lazdunski M, Frelin C (1989) The inotropic effect of endothelin-1 on atria involves hydrolysis of phosphatidylinositol. Biochem Biophys Res Commun 249:143–146

Wallace JL, Cirino G, de Nucci G, McKnight W, MacNaughton WK (1988) Endothelin has potent ulcerogenic and vasoconstrictor actions in the stomach. Brit J Pharmacol 95:C127

Wallnofer A, Weir S, Rüegg U, Cauvin C (1989) The mechanism of action of endothelin-1 as compared with other agonists in vascular smooth muscle. J Cardiovasc Pharmacol 13 (Suppl 5):S23–S31

Warner T, de Nucci G, Vane JR (1988) Release of EDRF by endothelin in the rat isolated perfused mesentery. Brit J Pharmacol 95:723P

Warner TD, Mitchell JA, de Nucci G, Vane JR (1989) Endothelin-1 and endothelin-3 release EDRF from isolated perfused arterial vessels of the rat and rabbit. J Cardiovasc Pharmacol 13 (Suppl 5):S85–S88

Watanabe T, Kusomoto K, Kitayoshi T, Shimamoto, N (1989a) Positive inotropic and vasoconstrictive effects of endothelin-1 in in vivo and in vitro experiments: characteristics and the role of the L-type calcium channels. J Cardiovasc Pharmacol 13 (Suppl 5):S108–S111

Watanabe T, Miyazaki H, Kondoh M, Masuda Y, Kimura S, Yanagisawa M, Masaki T, Murakami K (1989b) Two distinct types of endothelin receptors are present on chick cardiac membranes. Biochem Biophys Res Commun 161:1252–1259

Watanabe T, Suzuki N, Shimamoto N, Fujino M, Imada A (1990) Endothelin in myocardial infarction. Nature (Lond) 344:114

Webb DJ, Cockcroft JR (1989) Plasma immunoreactive endothelin in uraemia. Lancet i:1211

Weiser E, Wollberg Z, Kochva E, Lee SY (1984) Cardiotoxic effects of the venom of the burrowing asp, *Atractaspis engaddensis* (Atractaspididae, Ophidia). Toxicon 22:767–772

Winquist RJ, Scott AL, Vlasuk GP (1989a) Enhanced release of atrial natriuretic factor by endothelin in atria from hypertensive rats. Hypertension 14:111–114

Winquist RJ, Bunting PB, Garsky V, Lumma PK, Schofield TL (1989b) Prominent depressor response to endothelin in spontaneously hypertensive rats. Eur J Pharmacol 163:199–203

Wright CE, Fozard JR (1988) Regional vasodilation is a prominent feature of the haemodynamic response to endothelin in anaesthetized, spontaneously hypertensive rats. Eur J Pharmacol 155:201–203

Yamasaki H, Niwa M, Yamashita K, Kataoka Y, Shigematsu K, Hashiba K, Ozaki M (1989) Specific ^{125}Iendothelin-1 binding sites in the rat atrioventicular node of the porcine heart. Eur J Pharmacol 168:247–250

Yanagisawa M, Kurihara H, Kimura S, Tomobe Y, Kobayashi M, Mitsui Y, Yazaki Y, Goto K, Masaki T (1988a) A novel potent constrictor peptide produced by vascular endothelial cells. Nature (Lond) 332:411–415

Yanagisawa M, Inoue A, Ishikawa T, Kasuya Y, Kimura S, Kumagaye S-I, Nakajima K, Watanabe TX, Sakakibara S, Goto K, Masaki T (1988b) Primary structure, synthesis and biological activity of rat endothelin, an endothelial derived vasoconstrictor peptide. Proc Natl Acad Sci (USA) 85:6964–6967

Yanagisawa M, Masaki T (1989a) Endothelin, a novel endothelium - derived peptide. Biochem Pharmacol 38:1877–1883

Yanagisawa M, Masaki T (1989b) Molecular biology and biochemistry of the endothelins. Trends Pharmacol Sci 10: 374–378

Yoshimi H, Hirata Y, Fukuda Y, Kawano Y, Emori T, Kuramochi M, Omae T, Marumo F (1989) Regional distribution of immunoreactive endothelin in rats. Peptides 10:805–808

Yoshizawa T, Kimura S, Kanazawa I, Uchiyama Y, Yanagisawa M, Masaki T (1989a) Endothelin localizes in the dorsal horn and acts on the spinal neurones: possible involvement of dihydropyridine-sensitive calcium channels and substance P release. Neuroscience Lett 102:179–184

Yoshizawa T, Kimura S, Kanazawa I, Yanagisawa M, Masaki T (1989b) Endothelin-1 depolarizes a ventral root potential in the newborn rat spinal cord. J Cardiovasc Pharmacol 13 (Suppl 5):S216–S217

Yoshizawa T, Shinmi O, Giaid A, Yanagisawa M, Gibson SJ, Kimura S, Uchiyama Y, Polak JM, Masaki T, Kanazawa I (1990) Endothelin:a novel peptide in the posterior pituitary system. Science 247:462–464

Yoshizumi M, Kurihara H, Morita T, Yamashita T, Oh-hashi Y, Sugiyama T, Takaku F, Yanagisawa M, Masaki T, Yazaki Y (1990) Interleukin-1 increases the production of endothelin-1 by cultured endothelial cells. Biochem Biophys Res Commun 166: 324–329

Yoshizumi M, Kurihara H, Sugiyama T, Takaku F, Yanagisawa M, Masaki T, Yazaki Y (1989) Hemodynamic shear stress stimulates endothelin production by cultured endothelial cells. Biochem Biophys Res Commun 161:859–864

Zeidel ML, Brady HR, Kone BC, Gullans SR, Brenner BM (1989) Endothelin: a peptide inhibitor of Na^+K^+-ATPase in intact renal tubular epithelial cells. Am J Physiol 257:C1101–1107

Subject index